Mining Capital

Michael Seeger

Mining Capital

Methods, Best-Practices and Case Studies for Financing Mining Projects

Springer

Michael Seeger
Munich, Germany

ISBN 978-3-030-31224-4 ISBN 978-3-030-31225-1 (eBook)
https://doi.org/10.1007/978-3-030-31225-1

Cover illustration with kind permission by Stefanutti Stocks Mining Services

This Springer imprint is published by the registered company Springer Nature Switzerland AG.
The registered company address is: Gewerbestrasse 11, 6330 Cham, Switzerland

I dedicate this book to my wife Glenda and my children Karsten, Keenia, and Kessia and their future children.

Preface

Mining capital is capital required to develop mines from exploration, feasibility studies, mine design, construction, and commissioning to operation; capital to refinance and modernize old mines; capital to purchase mining, processing, and related mine infrastructure equipment; and capital to acquire mines.

The supply of commodities remains vital to the development of modern society. While the minerals in demand have moved toward metals required for electric vehicles, renewable energy technologies, and smart phones, base metals and industrial minerals remain vital for the construction industry and manufacturing industry. A growing population requires food stability, and fertilizers produced from minerals remain vital. Energy minerals ensure that base load power is available for energy generation together with renewable energy. And in a turbulent world, gold retains its value and purpose of being the ultimate value foundation.

The bulk of the commodities need to be mined by existing and new mines. These need to be developed, financed, and managed by professional mining teams that operate mines in an environmentally and socially sustainable manner.

This book is about financing of mines, i.e., the raising of mining capital. Mining capital involves large sums of money and risk taking from both the investor and the mine developer. The mine developer risks time, start-up capital, and reputation while developing the mining project, whereas the investor risks losing the investment.

Therefore, a strong mining business case is fundamental to mining capital. Only a mining business case that is economic from orebody to market and that can hold in the face of many variables in mining, such as changing geology, moving commodity markets, technological changes, increasing environmental social pressures, and constantly changing host country dynamics, can and should be funded. The mining business case covers the geology, mining, processing, the market, the logistics path, the team, all licenses, and the environmental and social sustainability of the mining venture, the production rate and revenue model, capital cost, and operating costs. The mining business case is summarized in an economic model, and it is only as good as the assumptions made. This is where the team and its mining skills and experience play a very important role in the success of a mining business case.

The mining business case dictates the type of mining capital applicable for financing of mines. Many traditional forms of debt and equity finance for mining projects, disillusioned by the poor investment results of the past, have given way to new, alternative forms of capital, such as strategic partnerships with larger mining companies via farm-in agreements, reverse listing, development finance, streaming, royalty, and offtake finance from commodity traders intending to secure the supply of commodities, corporate bond finance, and finance from contractors intending to secure long-term mining and processing contracts via vendor finance or equipment finance.

This book provides a detailed description of each mining finance type and with it the term sheet and transaction structure in a manner that can be adapted by the reader for any mining business case in any commodity across the globe. Detailed case studies are provided for each mining finance type outlining the business case and funding plan behind diverse large, medium, and small mining transactions.

New miners must develop projects in more complex geological settings, often close to existing communities, in dry arid areas or in hostile political terrain. Environmental and social compliance is becoming more stringent and the era of easy mining is no longer there. The mining sector must compete for finance against other sectors with a similar risk profile such as biotech or fintech. Millennials do not see the mining sector as a first choice of employment, and modern mining skills are essential to keep developing and operating sustainable mines. The success of the mining sector for the future depends on the ability of mine developers to finance new mining ventures, and this book intends to provide mining finance skills to a new generation of mine professionals.

This is where mining capital comes in. It is the capital that unlocks the opportunity of developing and operating the mines of the future. This book will certainly enhance the success rate of financing mining projects and operations for any reader who studies and applies the principles outlined.

This guide is primarily intended for:

Mine developers seeking finance for their mining project:

This guide provides a practical and proven roadmap for mine developers to secure mining capital. The approach of building a mining business case, developing a funding plan, and selecting the type suitable to fund the business case as outlined in this guide will enhance the success rate to close a financial transaction. Case studies can serve as a reference for the different types of mining capital.

Investors intending to invest in mining projects and operations:

This guide covers the fundamentals of a solid mining business case that will attract capital and enable a return on investment. Mining investment entails investing large amounts of capital in a high-risk, high return environment which the investor needs to understand. This guide provides the structured approach to mining investment that any mining investor needs to consider, *prior* to making the investment.

Mining and processing contactors requiring equipment and plant to execute contracts:

In many cases, the mine owner outsources the mining and processing functions to EPC and OEM contractors, who must finance the mining and or processing plants equipment as part of their contract. This equipment must be financed by these contractors and will have to be paid off over the life of the contract. This guide provides a structure how to finance such equipment.

Host governments seeking to develop their mining sector:

In developing countries with natural resources, a strong mining sector that produces taxes can be the foundation of developing the economy with diverse industries. The fundamentals of a profitable and suitable mining sector are economic mineral resources, a business case for the mining projects to be developed, and mining capital. This guide can serve as a reference to governments seeking to develop their mining sector and attract investors to invest in their sector mining. The investor's perspective is highlighted.

About the Book

This book is intended to be a practical guide on how to successfully raise capital for mining projects. It is based on the accumulated experiences of the author, having experienced the successes and failures of raising capital for diverse types of mining projects, at different development stages in different jurisdictions. The book is intended to be used as a hands-on guide and covers the technical, financial, legal, and psychological ingredients that make investors part with large sums of money to finance mining projects.

The book is structured into five chapters that take the reader from the current characteristics of mining capital and dynamics affecting financing of mining projects to the process of developing a mining business case for investment, marketing the mining business case, raising mining capital, and case studies. The reader can read this guide from the beginning to the end or move straight to the relevant chapter.

Chapter 1: Characteristics of Mining Capital

Chapter 1 deals with the diverse applications of mining capital, provides a detailed analysis of the current dynamics affecting the success rate of financing mining projects and operations, and outlines recent mining deal examples. This chapter provides the context of the following chapters.

Chapter 2: Developing a Mining a Mining Business Case for Investment Methods

The ingredients of developing a mining business case are discussed: the process of developing a mine, the business case structure, the importance of a team, legal structuring, financial analysis, valuation, risk management, environmental and social sustainability, implementation plan vs. reality, and defendability. The mining business case is summarised in a project matrix.

Chapter 3: Marketing the Mining Business Case

The two key documents to market the mining business to investors are unpacked, in terms of structure, content, and case studies: the mining investor presentation and the mining business plan. The reader can adapt these to his or her own mining project or operation seeking capital.

Chapter 4: Raising Mining Capital—Best Practices

The reader is equipped with the tools to raise capital for mining projects successfully: the different mining capital types, capital for the different mine development phases, combining mining capital types for a successful financing, the process of due diligence for the mine developer and the investor, term sheets behind each mining capital type, the process of raising mining capital, the human dynamic, non-fundable mining deal proposals and how to rectify them, and a mining capital game plan that can be applied to any mining project, at any developments stage seeking capital.

Chapter 5: Mining Capital Case Studies

Case studies are presented as a practical reference to the reader, outlining both the mining business case behind the deal and the funding plan, with the mining capital type, term sheet, and transaction structure. The case studies involve transactions across the globe, in diverse commodities, at different project development stages for equity finance, farm-in finance, listing, reverse listing, project finance, development finance, corporate bond finance, streaming, royalty, and offtake finance, insurance wrapped finance, contractor vendor finance, equipment finance, and disinvestment finance.

Contents

About the Author

Michael Seeger BSc, MSc Eng, PhD, is a qualified mining engineer with extensive experience in mining finance, mine development, and mining operations in energy minerals, base metals, precious metals, and industrial minerals. Born in South Africa and raised in Germany, he served in the German Navy before studying mining engineering and mineral economics at the University of the Witwatersrand.

He started his career with Iscor Mining in coal, iron ore, and titanium mining operations and mine development. He was part of the part of the team that developed the titanium mining portfolio of Iscor Heavy Minerals, now owned by Tronox. He consulted for major gold and metals mining companies in Africa, South America, North America, and Europe.

He founded an opencast mining contracting company which mined thermal coal for Anglo Coal and was a cofounder of ASX listed Universal Coal, where he developed the coal project portfolio.

He is the founder of MX Mining Capital Advisors, which advises mine developers, mine owners, and mining investors on financing of mining projects and operations. The company is working on the financing of diverse mining projects in precious metals, energy minerals, base metals, and industrial minerals in Africa, Asia, the Americas, and Europe. He regularly speaks at major mining investment conferences on financing of mines.

He is married with three children.

List of Abbreviations

Mining Technical Abbreviations and Definitions[1]

ADT	Articulated dump truck
Ag	Silver
Au	Gold
CIL	Carbon in leach gold process
CIMVal	Canadian Institute of Mining, Metallurgy and Petroleum Property Valuation Committee
CPR	Competent Person's Report. A report on the technical aspects of a project or mine prepared by a Competent Person
cpht	Carats per hundred tons
CV	Calorific value measured in MJ/kg
DMS	Dense medium separation process
Doré bar	A semi-pure alloy of gold and silver created at the mine site
EIA	Environmental Impact Assessment
EMP	Environmental management programme
EPC Contractor	The contracting company that is appointed to engineer, procure, and construct the mine
EPCM	A contract specifying engineering, procurement, construction, and maintenance
FOR	Free on rail
FOT	Free on truck
g	Grams
JORC Code	The Australasian Code for Reporting of Exploration Results, Mineral Resources and Ore Reserves
kt	kilo tons (thousand tons)
LDV	Light duty vehicle
LHD	A load haul dump truck
LoM	Life-of-mine. The life of the mining operation, measured in years
LME	London Metals Exchange

[1]Source: the author based on compilation of various mining industry technical definitions.

m	Meters
mill	Million
Mt	Million tons
MW	Megawatt
O&M	Operate and maintain
P_2O_5	Phosphate
Resources	A concentration of minerals in such form, quality, and quantity that there are reasonable prospects for eventual economic extraction. Mineral resources are subdivided, in order of increasing geological confidence, into inferred, indicated, and measured categories
RC drilling	Reverse circulation drilling
Reserves	Mineral reserves are resources known to be economically feasible for extraction
ROM	Run of mine is ore mined and delivered to a processing plant
SAMREC Code	The South African Code for the Reporting of Exploration Results, Mineral Resources and Mineral Reserves
Sn	Tin
Stripping ratio	The ratio waste tons or m^3 to ore tons
t	Tons
tph	Tons per hour
tpa	Tons per annum
tpm	Tons per month
TiO_2	Titanium
V_2O_5	Vanadium
Works	All the technical and construction work required to develop and operate a mine

Mining Finance Abbreviations and Definitions

ASX	Australian Stock Exchange
Bond	A debt finance facility used by mining companies to raise capital for mining projects. The debt holder of the bond loans capital to the issuer of the bond, the mining company, for a defined period at a defined interest rate
BOO contract	A contract requiring the contractor to build, own, and operate the plant
BOOT contract	A contract requiring the contractor to build, own, operate, and transfer the plant to the client
CAPEX	Capital expenditure
C$	Canadian dollar
Capital structure	The mixture of capital (debt, equity, mezzanine) that the mining companies use to finance projects and operations

Condition precedent	The specific event required for a contract to commence
Convertible bond	A debt facility that can be converted into equity at the discretion of the investor
Corporate bond	Corporate bonds are issued by the mining company to the investor to raise capital, with payment obligations and collateral provided by the mining company
Coupon	Annual interest rate
Crowdfunding	Crowdfunding is a method of raising capital from individuals via an IT platform
Covenant	A specific business activity required
DSCR	Debt service cover ratio, defining the cash flow available to cover debt repayment
Debt finance	Debt finance is capital raised by a mining company from banks or financial institutions with the obligation to repay capital plus interest
Default event	An event that constitutes a contractual breech
DFI	Development finance institution
Equity finance	Equity finance involves a sale of shares in the company in exchange for capital
Escrow account	An account opened by a third party on behalf of other parties
Equator principles	A risk management framework used by financial institutions, for determining, assessing, and managing environmental and social risk in projects
EPCM	A contract specifying engineering, procurement, construction, and maintenance
Financial instrument	Cash, contract, or ownership of an asset
Facility	Loan
Guarantor	An entity assuring payment in an event of a default by the lender
Holdco	The holding company that holds the majority shares in the SPV
IPO	Initial public offering
IRR	Internal rate of return
Market capitalization	The total value of the mining company calculated by multiplying a company's shares with the market price of the share
Mezzanine finance	A hybrid of debt and equity finance, combining elements of both types of finance
LLCR	Loan life coverage ratio: The NPV divided by the debt owed by the company. It is used to determine the ability of the project to repay debt
Non-recourse loan	A loan which is secured by the project and its assets and which has no recourse to shareholders

NPV	Net present value
Obligor	Debtor
Off balance sheet financing	Financing which is kept of the balance sheet of the lender, typically a lease
OPEX	Operating expenditure
Option	A contract permitting the holder to buy or sell shares in a project at a predetermined price
Payback period	The amount of years it takes to repay a loan
PLC	A public limited company
Private placement	Raising of capital by selling shares to a few select investors
Political risk insurance	Insurance that insures project owners and investors against loss because of political events
Reserve tail ratio	Ratio of ore reserves to be mined at maturity of the finance facility to the total life of mine ore reserves
SPV	Special purpose vehicle
Senior debt	Debt that must be repaid first
Syndicate	A group of funders constituted and working collectively for a financing transaction
Sponsor	Owner of the project company
Standby facility	An additional loan facility to be used as a fallback facility
Subordinated debt	A junior loan subordinated to the senior loan
Surety	Guarantee of debt
Tenor	Duration of the loan in years
Vendor financing	Financing provided by the mining or processing contractor in the form of equipment or services which is repaid through operating costs which include a capital repayment portion
Waterfall payment	The payment schedule according to which creditors are being paid, with senior creditors being paid before lower ranking creditors
Wrap-up insurance	Insurance that protects the project owner, contractors, and funders

Characteristics of Mining Capital

1

Contents

1.1 Introduction to Mining Capital

Mining capital is the capital required to conduct mining projects. Capital is required to conduct exploration, feasibility studies, design mines, purchase mining equipment, processing plants, develop mine infrastructure or recapitalise producing mines.

It has become more challenging for mine owners and mine developers to raise money for mining projects and existing mines requiring capital since the downturn of commodity markets in 2010. Traditional sources of mining capital, debt and equity, have become more difficult to obtain, but at the same time alternative mining capital sources have evolved, enabling projects to be developed.

In the context of lower and fluctuating commodity prices, junior mining and exploration companies are battling to raise capital and multiple jobs in the mining sector have been shed. It is the classical bear market. There is a lack of conventional equity and debt capital and many investors shy away from investing in mining projects. It is a tough business environment for many mine developers and mining operations, as the bearish cycle takes its toll in the industry. Only larger mines with a strong balance sheet have easier access to conventional capital. However, this is part of the cyclical nature of the mining industry. A boom period, such as experienced in the 2010 is followed by a bear cycle, which is then followed again by a boom period. So, prepare for the next boom period.

M. Seeger, *Mining Capital*, https://doi.org/10.1007/978-3-030-31225-1_1

Whether boom or bust period, mine owners and developers need to raise capital for their projects or operations to remain in business. The type of the investor alternates between boom or bust periods. In a boom period, it is easy to gain access to mining capital as many investors want to invest in mining projects. In a bust period, investors become scarce. The mine developer needs to evaluate alternative ways of financing mining projects. In the bear market, cash flow-generating assets, or those with the promise of moving into a position of generating cash flow, are in favour.

This guide shows how developers can secure mining capital, whether the industry is in a boom or bust phase. The mine developer faces the task of persuading the financier to transfer money into the bank account of the mine developer, so that the mining investment plan (we will term it the business case) can be realised. Accordingly, the mine developer needs to have an enticing investment proposal which needs to be packaged in such a manner that the investor becomes interested in the project. In exchange for this investment, the mining investor is issued with shares in the project or a guarantee of a return. As mining projects involve large sums of money, the process of due diligence and decision-making is time consuming.

Mining investment also involves risk-taking as mining has many variables, ranging from geology, to markets, host country dynamics, equipment input costs to manpower dynamics. Accordingly, mining investors seek a high return on investment.

1.2 Mining Capital Applications

The mining industry is a capital-intensive industry and there is a constant demand for mining capital. The parties requiring mining capital are categorised as follows:

Capital to Advance Exploration Projects
Exploration and junior mining companies require capital to conduct exploration campaigns, ranging from sampling, aeromag surveys, drilling campaigns to the development of geological models (Fig. 1.1). An exploration campaign can cost $100,000 to $5 million pending the size of the project.

Capital to Conduct Feasibility Studies and Design Mines
Mine developers need to conduct a feasibility study covering geology, mining method and mine planning, mineral processing, mine infrastructure, equipment and engineering, the market, the logistics path to the market, the manpower plan, and economic analysis, an implementation plan and a risk management plan (Fig. 1.2). The feasibility document conducted in-house or external can range from $0.3 million to $10 million.

Capital for Construction of Mining Projects
Once the feasibility study is approved, mine developers proceed with the implementation of the project following a process of engineering and design, procurement, land acquisition, site establishment, construction, and commissioning (Fig. 1.3).

Fig. 1.1 Exploration activities and drill core (Source: Gemecs)

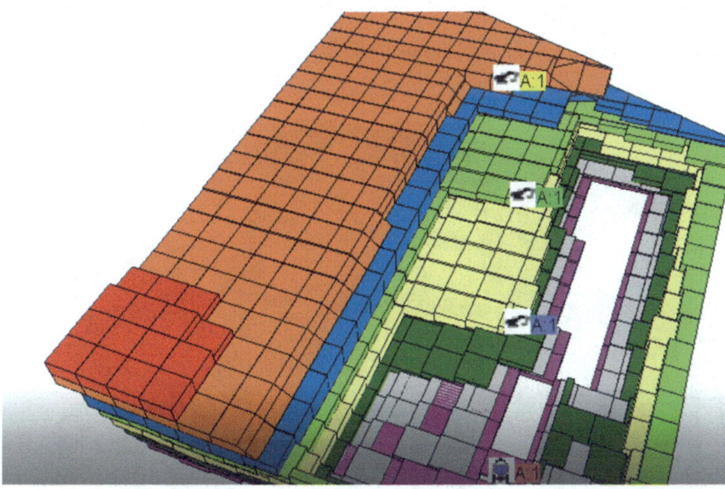

Fig. 1.2 Mine planning model for feasibility study (Source: The author)

This is the largest capital component and ranges from $5 million to $500 million pending the size and commodity of the project.

Capital to Purchase Mining and Processing Equipment

The primary equipment used on mines is mining equipment to drill, blast, load and haul the rock (Fig. 1.4), mineral processing plants to process the ore to produce mineral products (Fig. 1.5) and support equipment such as power plants. Modern mining projects are frequently structured such that the mine owner outsources the

Fig. 1.3 Mine construction (Source: Nautilus Projects)

Fig. 1.4 Opencast mining equipment (Source: GHM Sany)

mining and processing functions to EPC and OEM contractors, who vendor finance the mining and or processing plants equipment as part of their contract. Mining equipment ranges from \$0.3 million to \$50 million, pending the size of the fleet, whereas mineral processing plants rage from \$5 to \$50 million.

Fig. 1.5 Mineral processing plant (Source: MRD)

Fig. 1.6 Mining operation seeking capital for equipment refurbishment (Source: The author)

Capital to Recapitalise Existing Mining Operations

Once a mine is operational, the mining equipment, processing plant and mine infrastructure are subject to wear and tear. Mine owners are regularly required to raise capital to refurbish mine infrastructure, purchase new mining equipment, refurbish underground mine shafts or recapitalise and upgrade mineral processing plants with more modern technology. Recapitalisation can range upwards from $5 million to $50 million (Figs. 1.6 and 1.7).

Fig. 1.7 Mothballed underground mine seeking capital (Source: The author)

1.3 Mining Capital Industry Dynamics

The mining industry has seen significant changes and dynamics in the past 10 years, which all affect the type, nature and success of funding of mining projects and operations.

1.3.1 Mining Industry Dynamics

Lower, Fluctuating Commodity Prices with Slow Recovery

The period 2011–2018 has been a period of lower commodity prices as a result of a decrease in demand for metals and minerals by China. This has been followed by a slow recovery, with commodity prices often fluctuating. Lower commodity prices mean less supply of minerals and metals is required. This has led producers to focus on profitability of existing operations rather than new projects [1].

Host Governments Applying Pressure to Mining Companies

In many mineral-rich regions, host governments are increasing pressure on mining companies to increase community participation, increase the tax generated from mining and increase job creation. Mining companies are forced to work collaboratively with local governments and communities to meet these requirements [2].

Mineral Reserve Depletion

Over the past decade, the depressed commodity markets have led mining companies to cut exploration expenditure, cut capital expenditure for new projects and only focus on existing operations. With limited financing for exploration of pipelines, the global inventory of reserves has been depleting. Innovative financing models are required to kick-start exploration [2].

Quality Mining Projects Are Harder to Find

Good quality mineral deposits are harder to find. They are in difficult, remote places and the process of developing and financing the mine is more complex in such places. The easy mineral deposits have been found and mined. The increased difficulty and complexity increases the costs of exploration and mining [3].

Focus on Minerals for Future Technologies

Emerging technologies and the Green Revolution, driving renewable energy technologies, energy storage, electric motors and energy distribution and fuel require a select group of mineral and metals, for which new mines must be developed and financed (Table 1.1).

Disruptive Impact of the Digital Age on Mining Processes

The digital age and with it the process of automation of mining and mineral processes via robotic process automation and the use of data management to reassess exploration projects and optimise mining projects will have a disruptive effect on the mining industry. This means, in future, digital mining skills will be a key skill required to develop and operate mining operations [2].

Mining Technical Skills Retiring

Many skilled mining professionals with vast knowledge and experience, trained in different types of mining across the globe, have reached retirement age and are

Table 1.1 Green technology components and required minerals and metals

Green technology	Component	Minerals and metals
Hardware	Wind turbines	Silver
	Solar panels	Steel, silicon
	EV Chassis and Body	Aluminum, copper; silicon, steel
Energy storage	Lithium batteries	Lithium, cobalt, nickel, graphite
	Vanadium redox flow batteries	Aluminum, iron, copper, manganese, graphene, vanadium
Distribution and fuel	Nuclear fuel	Uranium
	Wiring Power grids	Aluminum, copper, silver, gold
Electric motors	AC induction	Copper aluminum, steel
	Permanent magnet	Neodymium

Source: Authors' table based on Vancouver Resource Investment Centre [4]

leaving the industry. The mining industry is looking to recruit young talent to make up for this shortfall and must compete with other more attractive industries to attract young people. There is a skills gap developing in the industry.

Water Scarcity a Factor for the Mining Business
Mining operations use significant amounts of water and are under increased scrutiny by communities and environmental groups to apply effective water management systems and conserve water. Dry mineral processing technologies are being developed to be applied in arid regions. The availability of water has become a factor in the ability to start mining operations and the financing of a mining business case [2].

Poor Historical Performing of Mining Investments
During the period of high commodity prices many mining projects failed to meet investor expectations. Mining companies focused on production growth rather than profitability. Budgets were missed and costs were inflated. In summary many mining investments did not compare with the returns of other investments, despite higher commodity prices [5].

1.3.2 Mining Industry Dynamics Affecting Financing Mining Projects

Subdued Investor Interest in Investing in Mining Projects
Many investors are no longer interested in projects with no cash flow and the absence of securable assets. Traditional sources of funding mining projects remain tight compared to the market peak in 2011, and many investors remain sceptical about the sector. The value and volume of mining deals have dropped (Fig. 1.8).

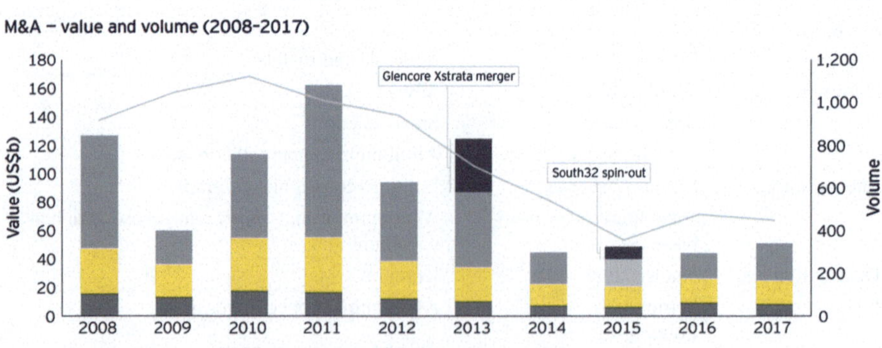

Fig. 1.8 Volume and value of mine deal sizes (2008–2017) (Source: EY's Global Mining and Metals) [6]

Mining Less than 1% of Global Project Finance

Decreasing commodity prices and a resultant shift of finance from mining into other investments have resulted in mining taking less than 1% of the global project finance [5].

Only Selected Projects Get Funded

In the business environment of constrained capital and lower commodity prices, mining companies are only pursuing the selected few, highest yielding projects to develop. The trend rests on profitability rather than expansion [5].

Exploration Spending Cutback

Spending on greenfield exploration projects has been cut back significantly and it has become challenging for explorers to raise equity finance. Historically, equity capital for exploration companies has been raised through listings. This route is no longer as easy in the current market as it was in the past [5].

Demand of Capital Outstrips Supply

It is easier for developers than explorers to raise capital. However, demand for capital outstrips supply and the cost of capital. Many traditional investors have become sceptical of investment in the resources sector at lower commodity prices [5]. This changes when commodity prices surge.

Listing Success Depends on Commodity Prices

During the feasibility stage, projects are still dependent on equity capital. In periods of lower commodity prices, several listings were cancelled due to poor trading volumes and high regulatory costs [5]. With the rising demand for battery technology metals, listings have become attractive again to raise capital for mines entering production.

Share Price Trend of Listed Junior Miners: Growth and Decline, Dilution for Shareholders

Many listed junior mining companies follow the trend of a rapid growth within a year of listing, followed by a period of sustained decline and levelling when expectations set by mine developers were not met. As project development results get announced, the share price moves upwards, followed by a decline as it is business as usual. Any raising of large capital dilutes existing shareholders substantially and innovative non-equity financing techniques are sought [3] (Fig. 1.9).

Retrenchment of Resources Teams in the Banking Sector

A trend has been the retrenchments of resources teams in the banking sector. This means less mining finance skill are available to assess mining investment opportunities.

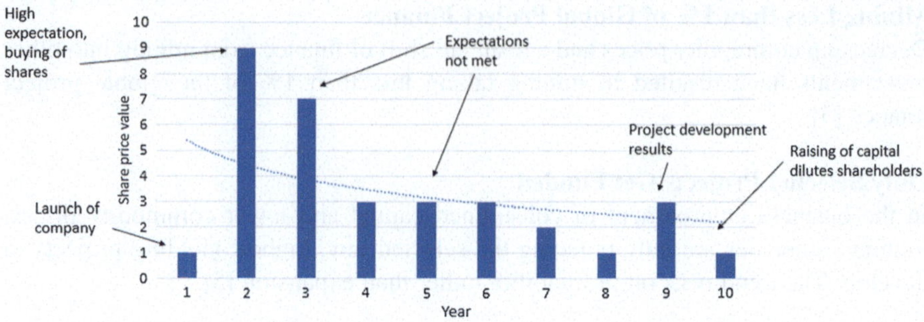

Fig. 1.9 Share price trend of junior mining companies affecting financing (Source: The author)

Capital Seeking Quick Returns Not Suitable for Mining
Modern capital seeks a quick turnaround. Mining projects are long-term investments and is unsuitable for shorter-term financing. Financiers must recognise this and short-term return expectations by modern capital providers can lead to bad experiences in mining finance.

Mining Companies Overleveraged
During the boom period, mining companies secured debt and share price appreciates. The downturn of commodity prices has resulted in a drop in share prices and lower market capitalisation of mining companies, while the debt obligation remains constant. This has resulted in many overleveraged mining companies.

Coal Mining Projects Difficult to Finance
Coal and coal-based power is seen as a major contributor to climate change. So much so that many banks no longer finance coal projects and major mining companies are divesting from coal. The demand for coal remains high as it is the base load power for many developing countries. Mining capital to the coal mining industry must be structured innovatively in the absence of traditional finance. Coal miners must focus on implementing green coal technologies, and advanced coal beneficiation such as coal to liquids or the production of fertilisers from coal to remain attractive to the investment community [7].

Smaller Mining Projects Easier to Finance than Large Mining Projects
Securing of capital for smaller high-grade projects is easier than for large mining projects. Bulk mineral projects (e.g. iron ore and coal) in developing countries require the construction of significant mine infrastructure and logistics capacity. This differs significantly from the CAPEX required to develop a small gold mine. In the current market, funding of large mining projects has been difficult and only the most attractive projects have been able to secure finance. Developers have downsized their projects to accommodate the market of lower commodity prices and challenging financing. Smaller projects can be financed more easily and scaled up once market conditions improve [8].

Redeployment of Mining Capital to Existing Operations
In an era of lower commodity prices, mining companies have allocated capital to existing operations rather than to new projects. The building of production capacity has become secondary, with the focus being the reduction of debt, dividend payments to shareholders, and where applicable, share buyback programmes.

Producers Have Easier Access to Mining Capital
Producers have more funding options available than projects in an exploration, feasibility or development stage. Producers have the advantage of cash generating operations, track records and they can secure loans, providing greater comfort to lenders. Raising capital remains easiest for mining companies with an investment credit rating [5].

Capital Varies for Different Types of Commodities
Mining capital varies for different types of commodities. Gold projects require a different type and sources of capital to base metals or energy mineral projects.

Funding Ease Dependent on Commodity
Funding is easier for mining projects producing minerals for the new economy, such as vanadium, lithium, titanium or phosphates. On the other hand, most banks do not fund coal projects anymore, as they are considered to contribute to global warming, and such projects require alternative funding methods. Mining capital available is commodity specific.

Creative Alternative Financing
Mining companies are seeking alternative and creative ways to finance the development of their projects from exploration to production, as a result of the difficulties of securing traditional sources of mining capital [9]. Alternative forms of financing that were less popular during the boom times have become important for miners of all stages. These are farm-in agreements and strategic partnerships, royalty and streaming deals, offtake finance, strategic partnerships, convertible bonds, EPC vendor financing and joint venture partnerships with traders and Chinese state-owned entities.

The downside of new alternative financing such as royalty, streaming and offtake financing is that miners sell their production forward at a discount to current commodity prices, forfeiting any upside in the commodity cycle. The different pre-2010 and post-2010 mining financing sources are illustrated in Fig. 1.10.

A comparison of alternative mining financing methods shows that highest growth is expected in royalty and streaming finance, followed by private equity (Fig. 1.11).

Commodity Traders with Capital Seeking to Secure Supply Chain
International commodity traders in precious metals, base metals and energy minerals seeking to secure their supply chain of commodities have become a source of capital and investors in mining projects and operations.

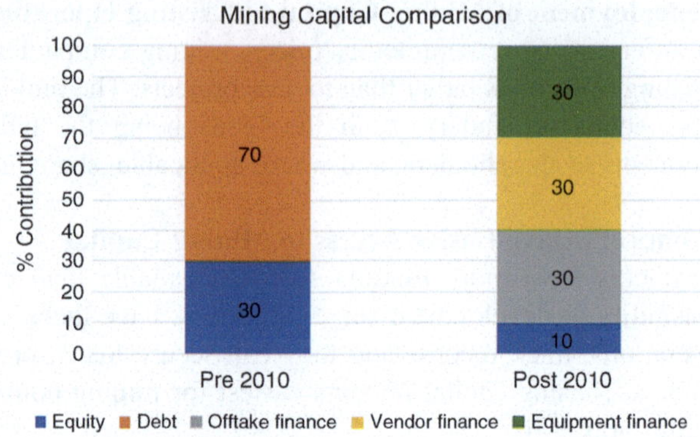

Fig. 1.10 Traditional vs alternative mine financing (Source: The author)

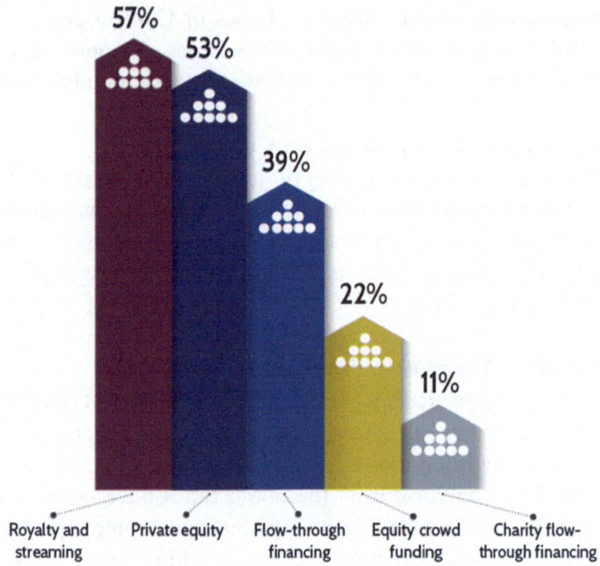

Fig. 1.11 Comparison of alternative financing methods (Source: The Northern Miner—New Frontiers in Mining Finance) [1]

Outsourcing and Vendor Financing
Mining companies have changed the operating models to suit the leaner capital environment. This means that the key function in the mining process is outsourced and vendor financed by specific suppliers, who recover capital via operating costs. This includes mining equipment, mineral processing plants, power plants, water treatment plants and other related mining infrastructure. Fundamental to this is a profitable mining business case [10].

Mining Companies Turn to Development Finance and Political Risk Insurance
Mining companies have turned to regional development finance institutions and political risk insurance to finance mining projects in host countries with higher political risks, when conventional financiers have avoided such countries.

Modular Implementation

Modern off-the-shelf mineral processing plant technology and designs enable mining projects to grow in an incremental, modular manner as opposed to a large expansion. Additional mineral plant circuits can be added to existing plant circuits which reduces the CAPEX requirements of projects [10].

Collaboration

As a result of lack of conventional capital, mining project developers and mine owners are collaborating in order to attract capital. This entails using mines sharing a central processing plant and related infrastructure or pooling a portfolio of exploration projects.

Disinvestment to Raise Capital

The sale of mining assets in order to obtain capital for development of another mine asset has become the strategy of mine owners and developers.

Distressed Junior Mining Sector Seeking Strategic Investors and Partnerships

The junior mining sector has been distressed for many years. The period of obtaining easy capital and drilling a few holes to attract the buyout by a larger mining company is no longer there and most juniors are starved of capital. With capital allocation to juniors being sporadic, the solution lies in junior miners entering into farm-in agreements with strategic investors as a source of capital to develop opportunities [11]. These are mid-tier or major mining companies seeking investment opportunities.

The Importance of Millennials to Mining Finance

Most millennials are not attracted to the mining industry and the ageing baby boomer generation is conserving capital and unwilling to risk it in the speculative mining ventures. The task of the mining industry is to persuade millennials that mining remains an important part of the economy. The mining industry is beginning to recognise that millennials are key to providing capital for mining projects in the future. The mining sector should be seen as to have the same risk/reward ratio than the biotech sector [1, 11].

Crowdfunding as Potential New Source of Finance for Exploration

Crowdfunding is a low-cost method of raising capital for speculative projects through pooling of capital from individuals via an IT platform. Early stage exploration projects in future metals are suitable for crowdfunding and this type financing may play an important role in financing of future exploration projects [1].

The current mining capital industry dynamics are summarised in Fig. 1.12.

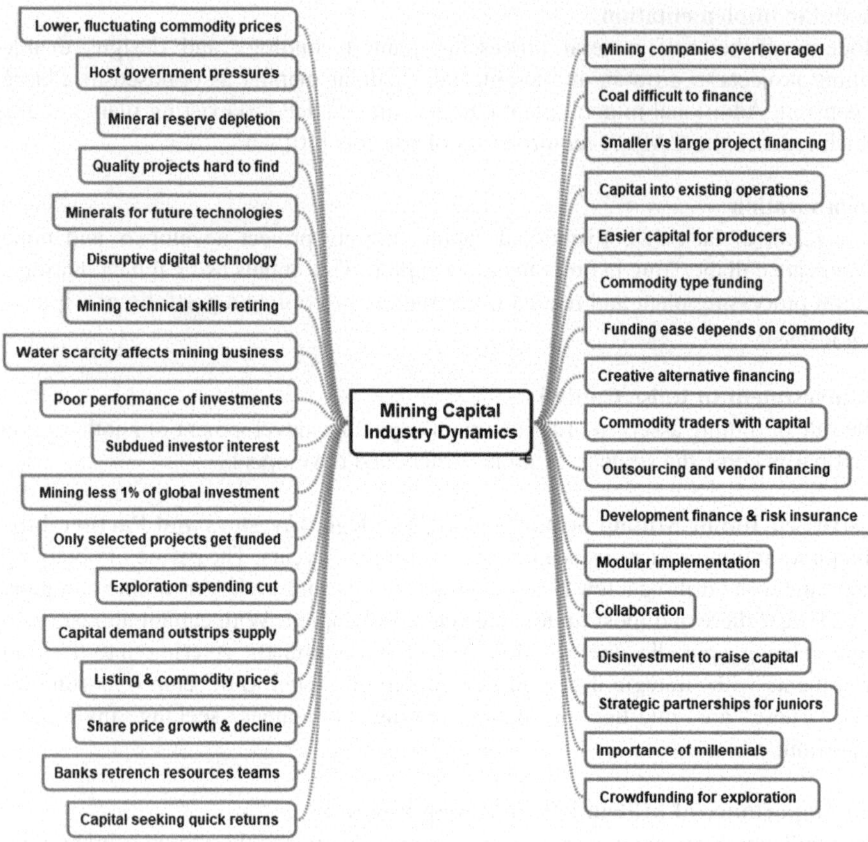

Fig. 1.12 Mining capital industry dynamics (Source: The author)

1.4 Mining Capital Deal Examples

Select mining capital deal examples of the period 2018–2019 are outlined in Table 1.2. The type of transactions involves debt finance, equity investment via share placement on stock exchanges, offtake finance, IPO, equity investment, development finance, convertible bond and vendor finance.

Table 1.2 Mining capital deal examples

Owner	Project	Country	Financier	Financial instrument	Deal size $M
Drummond Company	Shoal Creek metallurgical coal mine [12]	US	Peabody Energy	Acquisition finance	400
Randgold Resources	Randgold Resources acquisition [13]	Canada	Barrick Gold	Acquisition finance	18,300
Exxaro	North Block project [14]	South Africa	Universal Coal	Acquisition finance	13
Artemis Resources	Carlow Castle gold/copper/cobalt project [15]	Australia	RiverFort	Convertible debt	4
Millennium Minerals	Nullagine project expansion [16]	Australia	Investec	Debt facility	12
Nickel Mines	Ranger nickel project [17]	Indonesia	Decent Investments	Debt facility	80
Lion One Metals	Tuvatu gold project [18]	Fiji	Sinosteel	Debt facility	40
Edenville Energy	Zambia–Tanzania–Kenya power transmission line [19]	Kenya	World Bank	Debt facility	455
Ivanhoe	Kamoa-Kakula, Platreef and Kipushi projects [20]	South Africa	CITIC Metal	Debt facility	100
Ncondezi	Restructure of $5.1-million loan [21]	Mozambique	Strategic partners	Debt facility	5
Lucapa	Mothae kimberlite mine [22]	Lesotho	IDC	Development finance	7
Newmont Goldcorp	Newmont Goldcorp's Eleonore mine facilities [23]	Canada	Government of Canada's Québec	Development finance	1
Emmerson	Khemisset Potash Project [24]	Morocco	ASX	Equity/share issue	8
Poseidon Nickel	Silver Swan and Black Swan mine [25]	Australia	ASX Squadron ResourcesBlack Mountain Metal	Equity/share issue	53
Mineral Resources	Wodgina Lithium Project [26]	Australia	Albemarle	Equity acquisition	1150

(continued)

Table 1.2 (continued)

Owner	Project	Country	Financier	Financial instrument	Deal size $M
Reunion Gold	Gold exploration project Guiana Shield gold district [27]	Asia	Barrick Gold	Equity acquisition	2
Sun Metals	Company acquisition—Sunmetals [28]	Canada	Teck Resources	Equity investment	4
Centaurus Metal	Conquista exploration project [29]	Brazil		Equity investment	0.1
Mustang Resources	Mustang's Montepuez ruby and Caula vanadium/ graphite projects [30]	USA	US investor Arena Investors	Equity: Share placement	14
Talga Resource	Lithium-ion battery graphite and graphene product development [31]	Australia	ASX	Equity: Share placement	6
Peel	Wagga Tank-Southern Nights Project [32]	Australia	ASX	Equity: Share placement	5
Galilee Energy	Glenaras gas project [33]	Australia	Institutional and sophisticated investor	Equity: Share placement	4
Venture Minerals	Riley Project [34]	Tasmania	ASX	Equity: Share placement	4
SQM	Cauchari-Olaroz lithium project [35]	Argentina	Ganfeng	Equity: Share sale	88
Titan Minerals	Acquisition of the Mirador plant [36]	Peru	ASX	Equity: Share placement	8
Stonewall Resources	Theta Hill gold project [37]	South Africa	ASX	Equity: Share placement	4
Fortescue	Queens operation [38]	Australia	Inhouse	In-house financing	287
Minergy	Masama coal mine [39]	Botswana	AIM	Listing	2
Coronado	3 × US mines and Curragh mine in Australia [40]	US	Energy and Minerals Group	Listing IPO	910

Thiess	Sonoma North, Drake and Jax mines [41]	Australia	QCoal Cimic	Mining contract	336
Bauba Platinum	Moeijelijk mine [42]	South Africa	Gerald Metals and Pelagic Resources	Offtake finance	5
Lonmin	Company investment [43]	South Africa	Pangaea Investment	Streaming finance	200

Source: Authors' table based on review of mining transactions [12–43]

Developing a Mining Business Case for Investment: Methods

2

Contents

M. Seeger, *Mining Capital*, https://doi.org/10.1007/978-3-030-31225-1_2

2.1 A Strong Mining Business Case Will Attract Capital

In the current business environment of limited conventional mining capital and mainly sources of alternative financing, only a strong and solid mining business case will raise capital. Capital providers must be ruthless when analysing the business case of mining projects, as mining projects involve long lead times, high amounts of capital and involve risks that must be managed. When things go wrong in mining and large sums of money are lost, it is usually the business case that was flawed. When the mining business case is strong, large returns can be made for investors. Accordingly, this chapter equips the reader with the following tools to develop a strong mining business case for investment (Fig. 2.1).

- Development of a mine—from exploration to production
- The mining business case structure—from orebody to market
- Team—A strong team will convince investors
- Legal structuring—The mining business case needs a solid and simple legal structure

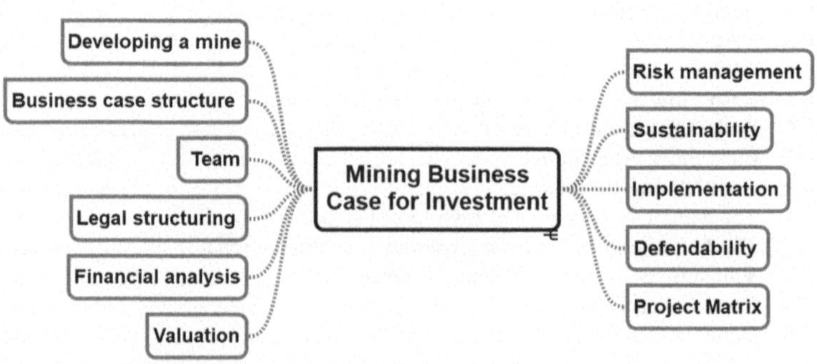

Fig. 2.1 Mining business case for investment—ingredients

- Financial analysis of the mining business case—the development of an economic model
- Valuation of mining projects—to enable negotiation on terms and conditions of the investment
- Risk management—Mining is risky business; risks need to be identified and managed
- Environmental and social sustainability—with a poor legacy, more than ever do mining projects need to be environmentally and socially compliant to attract capital
- Implementing the mining business case—plan vs. reality
- Defendability—Is the mining business case defendable?
- Summary—Mining business case Project Matrix

With a strong mining business case, the mine developer is ready to approach the market for capital.

2.2 Developing a Mine

2.2.1 Mine Development Steps

A mine developer seeking to raise mining capital needs to develop the mine in a structured and sequential process. This development approach is applicable to large, medium and small mining projects. Every mine project seeking to raise finance should be developed according to the following development stages [44]:

- Phase 1: Exploration
- Phase 2: Feasibility
- Phase 3: Mine design
- Phase 4: Construction and commissioning
- Phase 5: Operation
- Phase 6: Mine closure

It takes between 2 and 10 years to develop a mine, depending on the size of the operation. A mineral reserve is a finite natural resource. Once explored and proven to be economic, a mine is developed, and the mineral is extracted. When the mineral resources are depleted, the mining area must be rehabilitated (Fig. 2.2).

2.2.2 Phase 1: Exploration

Most mining ventures start out as a private group of geologists and mining professionals with an exploration concept. Financing is usually through an informal association of friends or venture capitalists who put up seed capital for exploration. Once the Prospecting Licence has been obtained, target generation commences, with

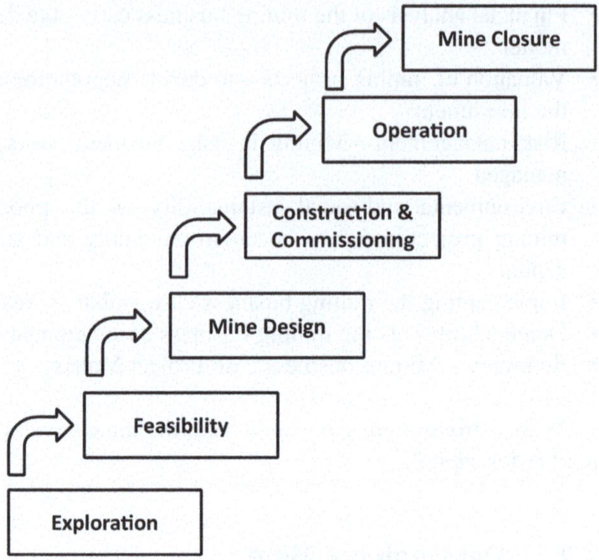

Fig. 2.2 Mine development steps (Source: The author)

a review of regional and project-specific technical information available on the project. The purpose of this exercise is to rank the projects and identify the most promising opportunities. Once the most prospective projects have been identified, the field survey can commence. This includes geological mapping, geochemical and geophysical surveys and trenching. Promising results will lead to a phased drilling campaign to define the resource. Drilling should provide sufficient information to classify the deposit as an inferred or indicated mineral resource. The data obtained would be incorporated into a geological model. Depending on the type and complexity of the deposit, this stage can take between 3 months and 2 years (Fig. 2.3).

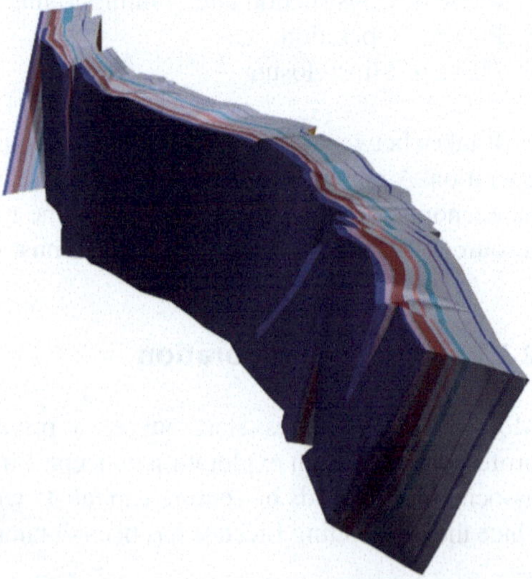

Fig. 2.3 Geological model (Source: Tivani Project, Ferrox)

2.2.3 Phase 2: Feasibility Stage

The feasibility stage commences once there are results from the exploration campaign. There are three phases to the feasibility stage of a project. They are:

- The concept study
- The prefeasibility study
- The bankable feasibility study

Concept Study
The concept study, often referred to as the scoping study, is a preliminary evaluation of the mining project based on general assumption and project input parameters of similar projects. The concept study identifies technical and economic parameters of a project, including target mineral reserve, the mining method, production targets, the most likely processing method, the logistical solution, potential target customers and an assessment of capital and operating costs. The level of accuracy is 50%. The concept study will determine if the project warrants further investigation.

Prefeasibility Study
The prefeasibility study involves further drilling in order to define the grade and tonnage of the mineral resource and further technical studies to evaluate the mining project. The basis for any further project development is a valid ore-reserve and geological information needs to be confirmed before any work can proceed. Mineral testwork has been completed to develop mine design parameters. The step following the establishment of the ore-reserve is the development of a production schedule, which takes into account the reserve tonnage and market demands. The production rate should be highly significant, as it determines the mine life, capital costs and operating costs of the project. A typical prefeasibility study should contain information and an analysis in the following areas to an accuracy of ±25%:

- Resources and reserves
- Mining method and production areas
- Processing method
- Mine infrastructure: water, power, site civils
- Mine waste management: discard dumps and tailings
- Mine logistics—mine to market
- Human resources planning
- Environmental and community considerations
- Capital and operating costs
- An economic analysis
- A risk management plan

The mining company will engage independent consultants to conduct preliminary studies in the above-mentioned fields. The tested method that investors use to analyse the potential success or failure of a project is the cash flow analysis that is summarised in a Net Present Value (NPV), Internal Rate of Return (IRR) and Payback Period. If the project is financially attractive, it has a positive NPV and meets the IRR hurdle rate set by the investor, which is usually greater than 20%.

Feasibility Study

The feasibility study is a refinement of all assumptions and design parameters of the prefeasibility study. This requires the technical and financial feasibility of the mine to be developed with a reduction of margin of error and is essential to support the decision to proceed with or abandon the project. Usually, the assumptions need to be $\pm 10\%$ correct.

The drilling campaign is extended to obtain more accurate geological information and mineral grades, which would enable the reserve to be classified as a measured mineral resource and increase investor confidence. Pilot plant trials are conducted on site to obtain technical data for mining and beneficiation. Specialist studies are undertaken to determine the social and environmental impact of the project. In-house and external consultants undertake further research in the fields of mining, beneficiation, marketing, infrastructure, transport and human sciences.

A critical issue during the feasibility stage is the political development in the host country. The host government only approves the required mineral licence if its requirements for local content, tax rates, royalties, and profit sharing are met. The payments by the mining company to the state for the project need to be agreed upon prior to the detailed design phase. Conversely, the company would not continue with the next phase if the political risk within the country were at an unacceptable level. In order to gain popular political support, the companies may give the state an equity stake in the project.

The NPV and IRR are recalculated based on the refined data. Sensitivity analyses are undertaken to determine the behaviour of the NPV and IRR with changes in the commodity price, production rates, revenues, costs, inflation and interest rates.

On completion of the studies, all the information is integrated into a bankable document. It is normal for the bankable feasibility document to be audited by external technical and financial consultants to confirm the overall conclusions. Once the audit has been completed, the mining company selects a gearing ratio and approach lending institutions to secure finance. Banking institutions employ in-house expertise to review and model project parameters to determine the risks associated with the project. In many instances, the company would also invite other companies to invest in the project to reduce its own risk exposure and balance its debt equity ratio.

2.2.4 Phase 3: Mine Design

Once the decision has been taken to proceed with the project, backed either by finance or with good prospects of obtaining finance, the mining company will undertake the detailed design of the project. Usually, consultants and the appointed

EPC and O&M contractors undertake the design, while specialists in the mining company review and approve the work.

Technical design

The mining method, based on the geology, the depth of the orebody, equipment availability, water and electricity and rainfall figures, must ensure that the deposit can be mined economically. The mining method and parameters selected have a direct impact on the cost of mining, the production rate as well as the ore recovery. The selection of the beneficiation process is based on small-scale beneficiation campaigns with in situ ore. Important issues are the milling method, the metallurgical recovery, the presence of oxidised material in the pit and relationships between tonnage, grade and recovery.

A logistics network must be established to ensure the proper transportation of product, people and equipment. The mine's infrastructure must be designed to suit the requirements of the project. To function properly, the mine requires the following infrastructure:

- In-pit crusher for opencast and hard rock mines
- Shafts, fans and underground infrastructure for underground mines
- Conveyor belts
- Beneficiation plant
- Electrical infrastructure, e.g. substations, powerlines, generators
- A water supply system
- Pollution control dams
- Access roads
- Workshops
- Rail or road loading facilities
- A railway line connecting onto the public railway line
- Tailings dam
- Administrative offices
- A sewage system
- Communication facilities
- A mining village, including schools, a clinic and recreational facilities

Certain infrastructure facilities may be available for mining projects near civilisation. However, developing countries and in remote areas, the available infrastructure might not meet the requirements for the efficient running of the operation. The mining company will then have to instal new infrastructure or upgrade infrastructure that is suitable for the project (Fig. 2.4).

Environmental and Social Management

Detailed studies on the social and environmental impact of the project must be undertaken and completed during the detailed design phase to ensure that the project meets the requirements of the host government, the local community and the mining investor. Mining companies have to submit an environmental management report

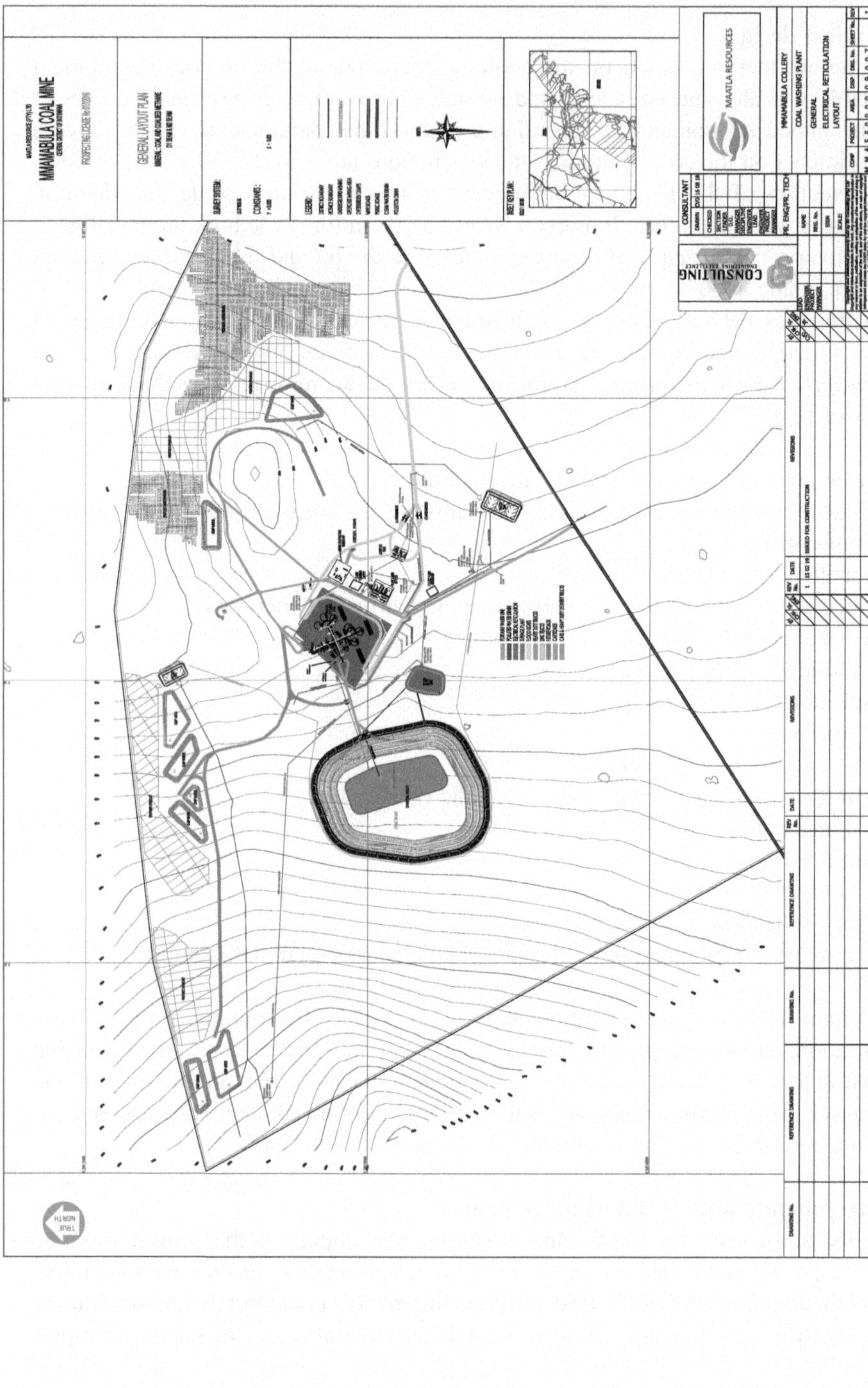

Fig. 2.4 Mine design (Source: Maatla Resources)

(EMPR) for approval before commencing with mining. Communication forums must be held on a regular basis to inform the local population of project developments.

Mining companies must place more and more emphasis on the environment and the social implications of their activities. Companies know that local and international pressure groups, indigenous people and the world's media are quick to act if the exploitation of minerals should lead to the exploitation of people, and by consequence, the environment. Financial institutions are increasingly concerned that projects should not only aspire to the environmental standards of the host country, but also to the highest international standards, as most countries were moving towards these standards at a rapid rate.

Human Resource Management

The mine developer must compile the mine organogram, job descriptions, salary specifications and training needs. Note, the management and personnel of associated with the mining project determine the success of the operation. Research must be done into the availability of local skills. Most of the employees should be locals and expert expatriates are only employed in key managerial and technical positions for a temporary duration.

Marketing

No project will go ahead without a market for the mineral commodity. Many projects fail to tie up the market causing unnecessary delays. Company marketing experts must continuously observe the commodity market in order to detect fluctuations in the supply and demand of their product. Long-term price volatility, cyclical behaviour and future price projections are generated for the project's commodity. Contacts with commodity traders are established and a substantial portion of the production is sold in medium- to long-term contracts or hedged to subject the project to a degree of stability. It is important for the project to enter the market at the upswing of the commodity price cycle in order to maximise revenue through increasing commodity prices.

Legal Issues

Various contracts must be concluded between the mining company and the associated partners. In a contract with the host government, royalties, import regulations and tax rates are clearly defined. Equipment suppliers, consultants and EPC construction companies are invited to tender for the construction of the mine. Because contractors can often provide services at cheaper rates than the mining companies, they enter legal agreements for the mining of the minerals, the transport of the product, engineering, process-related work and mine security.

Pre-production Activities

Prior to commissioning, pre-production activities are initiated. These activities include the recruitment and training of operational staff, the compilation of operational manuals, the establishment of bonus system and safety procedures, the implementation of information technology and the planning of the commissioning process.

2.2.5 Phase 4: Construction and Commission

Construction

The construction process commences once all CAPEX has been secured. The process commences with site clearance. This may entail the relocation and resettlement of affected villages and costly compensation. The construction of the mining infrastructure, surface infrastructure, tailings dam, beneficiation plants, roads, workshops, water supply system, electrical and communication infrastructure are completed as per a project schedule with stringent cost control and measurements and valuations of the works by an independent quantity surveyor (Fig. 2.5).

Commissioning

After the mine, the processing plant and the supporting infrastructure have been constructed, the commissioning phase commences. This can take up to 6 months, during which production is progressively increased to 100% capacity. Financial institutions undertake a completion test in which they observe the commissioning process to ensure that the mining company adheres to the design specifications.

2.2.6 Phase 5: Operation

Once the mine has been commissioned, the mine enters a normal production routine as per the approved production plan. The mining company will try to maximise the

Fig. 2.5 Mine construction activities (Source: Nautilus Projects)

return on investment through the effective management of the production process, people and equipment as well as by applying a philosophy of continuous improvement.

The typical operational process is as follows:

- The ore is mined and the Run of Mine (ROM) is transported to the mineral processing plant.
- At the mineral processioning plant the ore is beneficiated to produce a mineral product.
- The waste product generated from mineral processing is deposited into a slimes dam, waste dump or backfilled into the mining void.
- The mineral product is loaded by the logistics contractor and transported to the customer, via rail, road, sea or air.
- Equipment maintenance repairs are done on site in the workshop, while major overhauls require the importing of new equipment.
- Rehabilitation is undertaken concurrently with mining operation because the mining company usually has an obligation to return the mined-out areas to an acceptably rehabilitated condition.
- Royalties are payable to the host government. Tax might only be payable once the operation has achieved certain financial benchmarks or after an agreed upon date (Figs. 2.6, 2.7, and 2.8).

Fig. 2.6 Mining contractor operation (Source: Stefanutti Stocks Mining Services)

Fig. 2.7 Mineral processing operation (Source: MRD)

Fig. 2.8 Product handling and logistics (Source: The author)

2.2.7 Phase 6: Mine Closure

The mine closure phase will commence once the reserves are depleted. Normally this involves a scaling-down of operations and a phased retrenchment of the mine's

Fig. 2.9 Opencast mine rehabilitation (Source: The author)

employees. Where possible, employees are transferred to other operations or reskilled. Equipment is sold or transferred to other mines. Office blocks and workshops are demolished or taken over by the local community. The mining void and waste disposal site is rehabilitated and re-vegetated in accordance with the EMPR. Once the mine has been completely rehabilitated, a closure certificate is obtained (Fig. 2.9).

2.3 Mining Business Case Structure

2.3.1 Main Business Process: From Mineral in the Ground to Customer

The Mining Business Case covers the mineral resources and reserves, the mining, the mineral processing of the ROM material to generate a mineral product for market, the customer and the transport of the mineral product to the customer (Fig. 2.10).

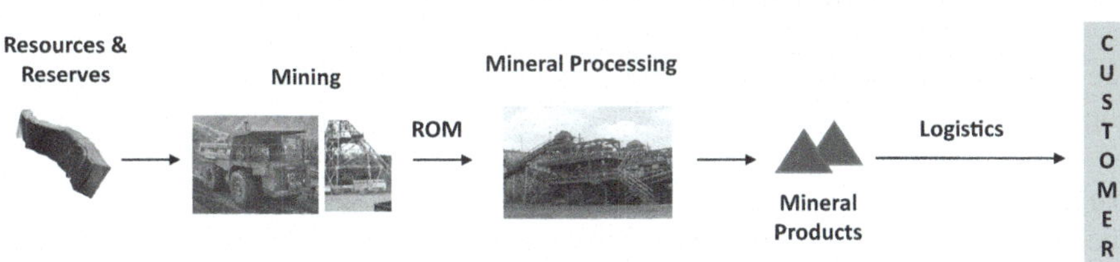

Fig. 2.10 Mining business case summary (Source: The author)

The mine developer needs a clear understanding of the Mining Business Case of his or her project to attract mining capital.

Resources and Reserves

The resources are the potentially extractable minerals in the ground.

The reserve is the economic mineable portion of a mineral resource. It considers depth below surface, cost of mining, environmental constraints, the revenue cost per ton of ore mined, capital and operating costs, geological and mining losses.

Mining

Mining entails extracting the economic portion of the using an open-pit or underground mining method. Open-pit mining usually involves a truck and shovel operation. There are diverse underground mining methods, such as board and pillar, stooping, block caving, vein mining, open stoping, block caving, scraper mining or long wall mining. Common to most mining methods is the process of drilling, blasting, loading and hauling.

Mineral Processing

This is the process of extracting and processing of minerals from the ROM ore through a series of mineral processes such as crushing, dense medium separation, magnetic separation and smelting. Depending on the mineral, the mineral process is complex or simple.

Logistics

This is the process of transporting the mineral product to the customer, either via road, rail, ship or plane.

Customer

The customer is the company that purchases the mineral product based on an offtake agreement with the mine. For base metal mining projects, this could be a smelter. For coal mines, this is coal-based power station. For precious metals, this is a gold or platinum refinery.

2.3.2 Activities in Support of the Mining Business Case

The activities in support of a strong mining business case are the following (Fig. 2.11):

Project Mission

The starting point of the mining business case is a project mission statement: the production of $\times$ million ton per annum by a predetermined date.

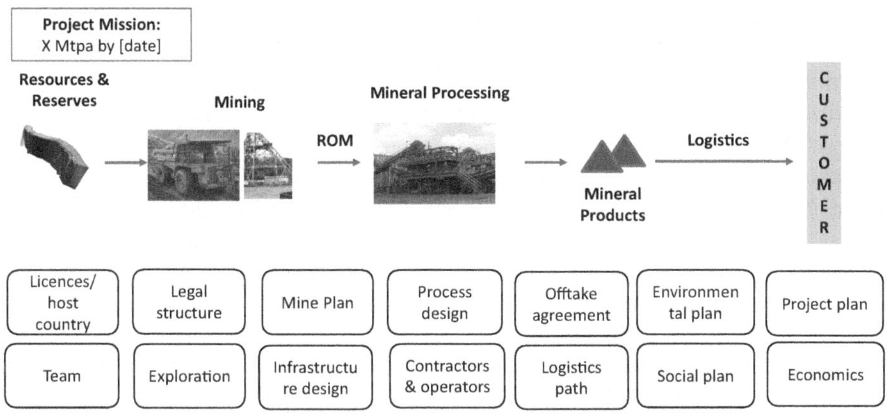

Fig. 2.11 Mining business case detail (Source: The author)

Licences/Host Country

Any mining project requires licences to allow the company to mine and extract the ore. The licences are the Prospecting Licence, the Mining Licence, the Water Licence, the Environmental Permit. The host country dynamics in which the project is located need to be considered.

Legal Structure

A mining project needs to be legally correctly structured. A Special Purpose Vehicle (SPV) has to be established to house the specific project. A holding company is usually the shareholder of SPV.

Team

A team with experience and a track record in developing mines is a key component to a successful funding campaign. The team will entail a geologist, mining engineers, process engineers, electrical and mechanical engineers, environmental and social scientists, project managers, lawyers, and cost controllers. The team may be contracted during the different stages of the project.

Exploration

An exploration campaign entails sampling, drilling, and development of a geological model. A thorough understanding of the orebody (tons, grade) is required to develop a mine plan for economic extraction.

Mine Plan

The mine plan is developed from the geological model. The mine plan has a yearly and monthly mining schedule, is based on a mining method suitable for the orebody and will provide capital and operating costs specific to mining.

Process Design
A process design entails the detailed design of the processing of the ore to produce the mineral product. The mineral processing plant is made up of several mineral circuits: crushing, screening, dense medium separation, leaching, smelting etc.

Infrastructure Design
Mine infrastructure entails the access road, access gate, weighbridge, civil and infrastructure design for the process plant, workshop, store, change house and mine offices, water and power supply to the mine, and the mine residue dam. The infrastructure design is conducted by civil, mechanical and electrical engineers.

Contractors and Operators
The mine processes—mining, processing and logistics will be operated by in-house staff or outsourced to mining, processing and logistics contractors. Contractors will have a defined scope of work and appointed in terms of a specific contract to execute the works.

Offtake Agreement
The mineral product is sold by the mining project to mineral traders. This requires a mineral sales agreement to be negotiated. The sales agreement will incorporate a production schedule, qualities, agreed ranges for production and quality targets, a term, sales price, escalation and penalties.

No market = no project go-ahead

Note: Often mine developers spend a significant amount of time and money to develop a project. Exploration work is completed, a mine plan is developed, a processing methodology is developed as well as the mine infrastructure plan. Assumptions are made in terms of the market. The company will sell the mineral product to market X based on the publicly known commodity market. The commodity price is included in the economic model. This approach works for precious metals. However, for base metals, industrial mineral and energy minerals it is vital to engage with the market early on. Mineral traders compete for product. The developer needs to secure an expression of interest on an official company letterhead that the mineral product will be bought. Price ranges and quantities need to be included. This gives the financier the required comfort that we can generate an income from the minerals produced. No market means that the project cannot proceed, and it is not financeable. Hence, as soon as possible developers need to engage with the market to avert this trap.

Note that a mining project developed in a depressed commodity market but launched during the upswing of commodity prices can be highly profitable.

Logistics Path
Transport of mineral product to the market: Mineral product is either sold FOT (Free on Truck) ex Gate, FOR (Free on Rail) or FOB (Free on Board). Logistics costs are often higher than the mining and processing costs combined. Shipping of the mineral

product is conducted by logistics companies engaged by mineral traders. The logistics costs are a key component in the economics of a mining project.

Environmental Plan
Any modern mining project needs to be environmentally compliant. The mining project needs to conduct environmental studies and develop mitigation plans to minimise the environmental impact of mining operations. This includes setting up a rehabilitation fund, which will be used to fund the mine closure.

Social Plan
New mines are near communities and need to be responsible neighbours. This means, the mining project needs to offer job opportunities to members of the community. Mining projects must compile a Social and Labour Plan that outlines concrete plans to develop the community through various initiatives such as the development of schools and technical centres to upskill the community.

Project Plan
The project plan must be developed from Prospecting Licence to completion of the mine and delivery of first ore to the customer. The task is to fast track project plan to approximately 3–4 years for midsized mining projects, implying that at the end of 4 years, the first product is delivered to the customer. Smaller mining projects can be established in less than 2 years. The project plan is a function of project economics, driven by market factors, CAPEX, OPEX, revenues as well as the geological confidence. The project plan is divided into activities and sub-activities, which are plotted against time: licensing, exploration, funding, feasibility studies, minerals marketing, sustainability, contracting, design, pre-production, construction, commissioning and operation. Key aspects of the project plan are resource allocation, ordering of long lead items, time to completion and ramp-up period.

Economics: NPV, IRR, Payback
Any mining project need to be analysed in an economic model. The economic model depicts on an annual and monthly basis the production and revenue, CAPEX, OPEX, royalties and a cash flow over the life of mine. Economic models can be simple or complex with multiple spreadsheets. The results of the economic model of a mining project are summed up in a Net Present Value calculated at a discount rate, an Internal Rate of Return (IRR) and a payback period.

2.4 Mining Team

2.4.1 Strong Mining Teams Attract Mining Capital

Mining investors look at the mining team, consisting of the mine developer and his/her supporting team members, when making an investment decision. A strong team, that can engineer and if required reengineer a project to make it fundable, will have higher chances of success at raising capital than a weaker team. A weak team can make a strong project unfundable. Key ingredients are mining and mining-related commercial skills and experience and the ability to function as a coherent team. The mining team is subject to intense scrutiny by prospective mining investors.

2.4.2 Members of Mining Teams and Their Roles

The members of the team are:

Mine Developer
The key player in raising mining capital is the mine developer or sponsor, as funders look particularly at the person(s) behind the SPV requesting finance. The mine developer needs to be an all-rounder with a thorough understanding of the specific mining sector in which the project seeking capital is positioned, have project development and operations management skills, and have commercial knowledge and acumen. Investors look at the track record of the mine developer in previous projects and operations and successful fundraising campaigns in the past. If funds were raised in the past by the developer, how did the investment perform?

Accountant
Accountant is the person developing financial models, budgets for CAPEX and OPEX, management accounts and project drawdown schedules.

Lawyer
The lawyer is responsible for concluding diverse agreements: shareholders agreements, lease agreements, acquisition agreements, service level agreements, mining and processing agreements, construction agreements, offtake agreements, royalty agreements and finance agreements.

Offtaker
An offtaker is the purchaser of the mineral commodity, the customer, usually a minerals trading company or end user of the mineral product.

Operator
The operator of the mine includes the mining contractor, mineral processing plant contractor and logistics contractor.

EPC Contractors
EPC contractors are the companies that are contracted by the mine developer to engineer, procure and construct the mine.

Consultants/Specialists
A mining venture requires different skill sets on an as-required-basis as the project evolves. Skill sets are diverse: legal, geological, geotechnical, mine planning,

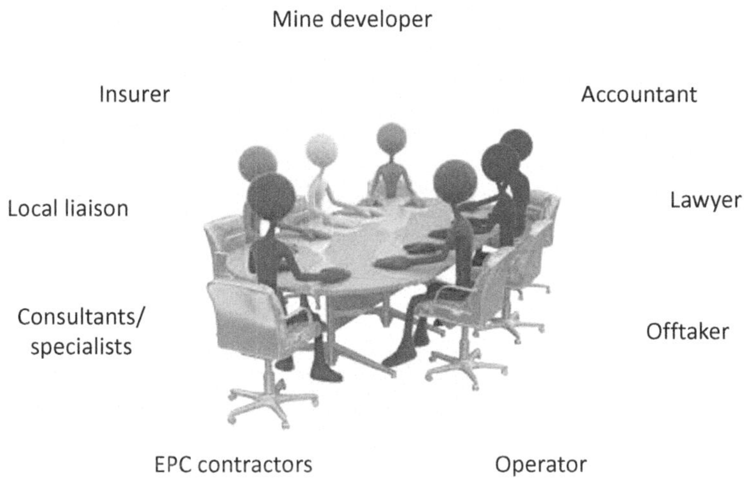

Fig. 2.12 Mining team members essential to success (Source: The author)

mineral processing, mechanical and electrical engineering, logistics, surveying, human resources, environmental, social and financial.

Local Liaison
This is the host country's representative in which the mining project or operation seeking capital is located, often termed local liaison director or officer, dealing with the regulatory, community and social aspects of the project.

Insurer
The insurer plays the key role of insuring the funder.

This team, with different sets of skills, expertise and networks, has to develop a game plan that delivers mining capital for a mining project or operation (Fig. 2.12). Not all members of the team are required to be part of the capital raise, but the key disciplines technical, market, commercial, legal and host country need to be covered.

2.5 Legal Structuring of Mining Deals

2.5.1 The Legal Structure of Mining Projects

The mining business case and the proposed mining deal need to be legally structured in such a manner that it is easy for the prospective investor to invest mining capital in the project or operation. A solid, simple legal structure improves the value of the business. A poor structure can lead to potential conflict in future or make it too cumbersome for the investor to invest. In most cases, a mining venture is structured as follows (Fig. 2.13):

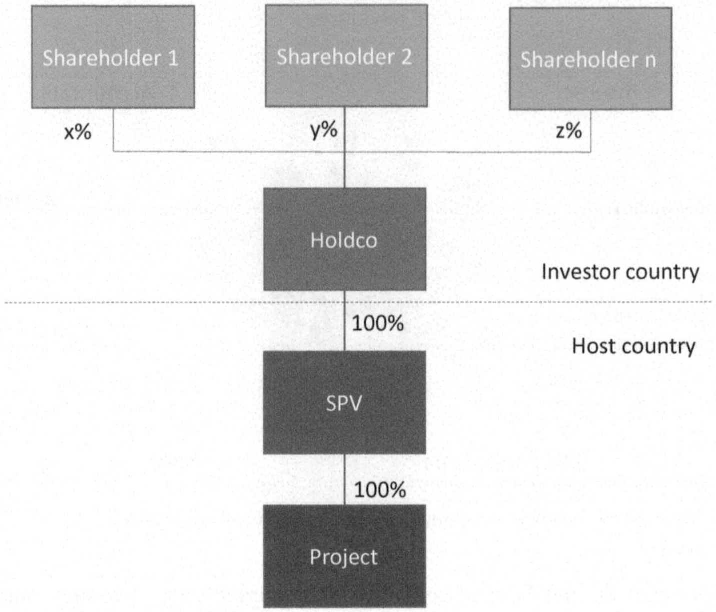

Fig. 2.13 Legal structure of mining deals (Source: The author)

Several shareholders form a holding company (Holdco), which could be an investment entity.

Holdco, in turn, owns the project company, termed Special Purpose Vehicle (SPV).

SPV owns the project rights, i.e. the prospecting licence, mining licence and related mining operation or mining project.

Holdco is often located in the investor/shareholder's country, while the SPV is in the host country where the project is located.

2.5.2 Mining Investor Legal Structure Preference

Investors like to invest in a simple structure. This is either at Holdco or at the SPV level. Investors usually want to invest as close as possible to the project level. A key criterion is repatriation of funds. The shareholders of the Holdco are in most cases the originators of the SPV and structure it to attract the investment.

Most investors do not want to invest in multiple layers above the project level. The structure depicted below shows the investor investing at a shareholder level, two layers above the project level. Some shareholders, if not aligned with all shareholders, tend to follow this route. This is also the route to follow if the investor wants to remain anonymous.

In general, this route adds to complication and is not preferred by many investors. The investor cannot exert as much influence as he could if he invested at a Holdco or SPV level.

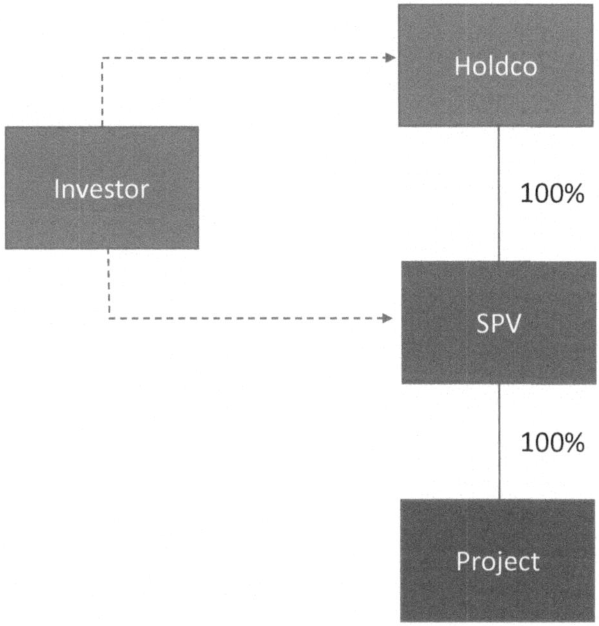

Fig. 2.14 Simple investment structure (Source: The author)

2.5.3 Incorporating State-Owned Entities and Communities in the Legal Structure

The structure below in Fig. 2.14 is common in host countries where legal obligations for a local company or trust exist, dictating that the community or the state holds a minimum stake in the project. Such a local company normally has a 'free carry', meaning it does not have to contribute to the CAPEX or OPEX (Fig. 2.15).

In this instance, the investor still invests at the Holdco or SPV levels. As Holdco and SPV have a balance sheet, they stand surety to the investor. The State/Community Company is often exempt of that obligation.

Such structures are debated extensively at a mineral policy level and a balancing act is required. Mining companies need to share the commercial benefits of the project with the host country/local community, while the structure needs to keep the prospective investor interested to invest in the host country (Fig. 2.16).

2.5.4 Structuring of Mining Deals

Mining deals are structured around the relationship of the SPV with the resources and reserves, the market, services providers, mining capital providers and surface rights holders (Fig. 2.17). The SPV as the owner of the project is at the centre of the

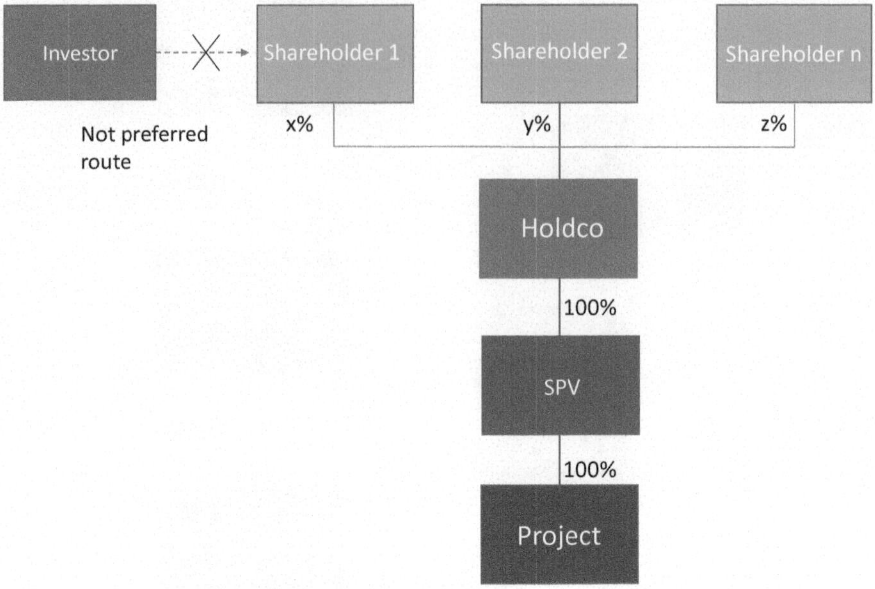

Fig. 2.15 Complicated investment structure (Source: The author)

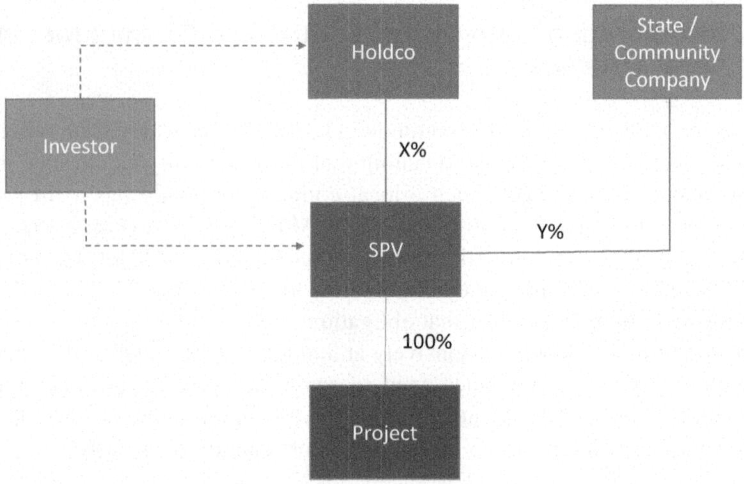

Fig. 2.16 Structure incorporating state-owned entities and communities (Source: The author)

value creation [45]. Service providers are either contracted externally, where most functions of the mine are outsourced, or work is executed in-house.

The generic contractual relationship of mining deals is typically as follows for mining projects seeking capital:

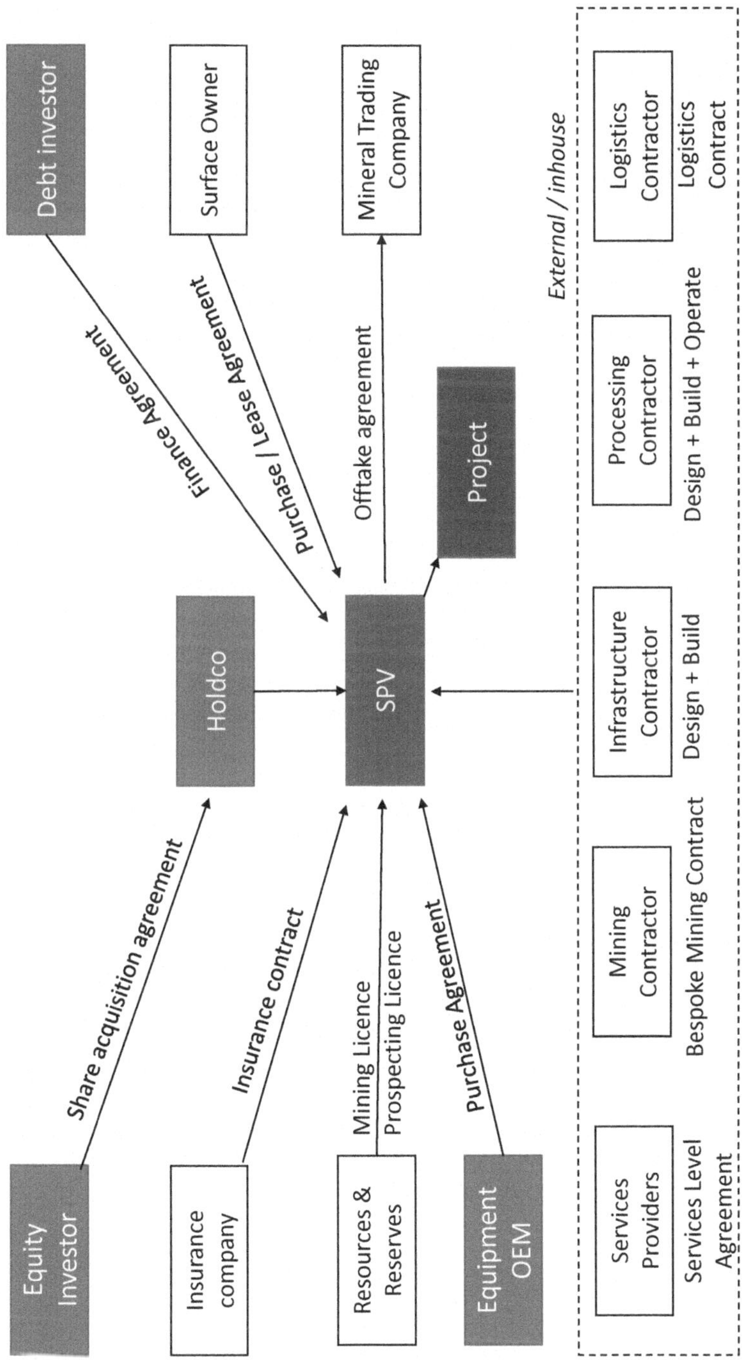

Fig. 2.17 Generic structure of mining deals (Source: The author)

Mining Licence The resources and reserves are secured by the SPV via a Mining Licence or Prospecting Licence.

Mineral Trading Company An offtake agreement specifying quantity, quality, pricing, delivery point, penalties, escalations, term for the minerals sales.

Surface Owner An agreement with the property owner, for the SPV to either purchase the property on which the mine is to be developed or conclude a surface rental agreement for the life of the project.

Mining Contractor A bespoke mining contract to mine the ore and waste and rehabilitate the mine by either opencast or underground mining methods.

Processing Contractor A mineral processing contract to process the ore. The typical format is a Design + Build + Operate Contract.

Infrastructure Developer A contract to Design + Build the mine infrastructure, including civils, water and power supply.

Logistics Contractor A logistics contract to transport the mineral to the delivery point, either by road, rail, ship or plane.

Service Providers Services agreement concluded between the SPV and technical, administrative and general service providers for the daily business of the mining operation or to development of the project.

Equity Investor A Share Acquisition Agreement between the shareholders of Holdco and the Equity Investor in exchange for investment in the SPV, which owns the Project.

Debt Finance Provider A finance agreement with SPV and the Holdco to provide finance at defined interest and repayment terms with securities provided by Holdco.

Equipment OEM A purchase or lease agreement with the SPV, mining or processing contractor to supply equipment for purpose of implementing the Project.

Insurer A business insurance agreement for the SPV's staff and personnel and against company loss and damage to property. The insurer can also insure the funder against losses and project non-performance.

2.6 Financial Analysis of Mining Projects

2.6.1 Purpose of Financial Analysis

The purpose of the financial analysis of mining projects is to determine the value of the investment opportunity. Investors want to understand project input parameters, CAPEX, OPEX, the project cash flow, and the profitability indicators.

The financial analysis entails the development of an economic model which is fundamental in mining investments and determines the value of the project equity.

There are simple and complex economic models. It is recommended to develop a simple economic model for issuance to the investors. The investor can then add additional features to the model.

The process of developing the economic model is outlined using the example of a gold project seeking mining capital for development.

Method of economic evaluation:

The method of developing the economic model follows a step-by-step process:

Step 1: Input parameters
Step 2: CAPEX
Step 3: OPEX
Step 4: Project cash flow
Step 5: Profitability indicators

2.6.2 Input Parameters

The input parameters need to outline critical input values and assumptions that the developer incorporates into the model. Each value needs to be supported by additional documentation. The developer needs to clearly reference the assumptions of each input parameter. This is where mining technical skills and experience come into play and investors will ask who is behind the data.

The input parameters are reserves, the production rate, the life of the mine, the mineral grade, the recovery, the commodity price, the commodity price, the CAPEX, OPEX, the tax rate and royalties. An example of the input parameters of a gold–silver mine is Table 2.1.

2.6.3 CAPEX

The capital expenditure (CAPEX) must be itemised, according to exploration, development costs, plant and equipment items and contingencies. The CAPEX list is as individual as each project is. The total CAPEX value is the quantum of investment that is required to further the project, the money that the developer from the investor in exchange for a return.

Table 2.1 Gold–silver mine input parameters for economic model

Parameter	UoM	#
Reserves	t	10,00,000
Production rate	tpa	60,000
Life of mine	Years	17
Au grade	g/t	3.6
Au recovery	%	85%
Au price	$/oz	1300
Au price	$/kg	40,625
Ag grade	g/t	210
Ag recovery	%	95%
Ag price	$/oz	17
Ag price	$/kg	531.25
Tax rate	%	28%
Royalties	%	2%

Source: The author

Table 2.2 Gold–silver mine CAPEX

Item	$
Licencing	2,50,000
Exploration	5,00,000
Mine design	1,00,000
Mine infrastructure	2,00,000
Mine equipment	22,50,000
Processing plant	64,00,000
Tailings disposal	4,00,000
Water and power	5,00,000
Vehicles	5,50,000
Rehabilitation	7,50,000
Contingencies	6,15,000
Working capital	10,00,000
Total CAPEX	1,35,15,000

Source: The author

The CAPEX list can be divided into development costs that have to be paid upfront and sustaining CAPEX that is distributed over the life of the project.

An example of a CAPEX is given below in Table 2.2.

2.6.4 OPEX

The operating costs (OPEX) of mining projects have the following categories:

- Mining
- Processing

Table 2.3 Gold–silver mine OPEX

Item	$/t
Mining	45.0
Processing	35.0
Logistics	8.0
Smelting and refining	26.0
Management	10.0
Environmental plan	2.0
Social plan	2.0
$/t	128.0

Source: The author

- Logistics
- Management
- Stay in business costs

The OPEX is depicted as a unit cost per sales ton.

The OPEX list must be supported by actual quotations and calculations.

In the simple model, the costs are not escalated, and are all real. A more sophisticated economic model will escalate OPEX prices. An example of a simple OPEX list for a gold mine is given below in Table 2.3.

2.6.5 Cash Flow

The project cash flow is calculated once the input parameters, the CAPEX and OPEX values have been determined by the developer. The project cash flow has the following categories:

- Production
- Revenue
- CAPEX
- OPEX
- Royalties
- Cash flow before tax
- Tax
- Cash flow after tax

The first 3 years of the cash flow sheet of a 10-year gold mining project are outlined in Table 2.4. The project starts at year 0, when the capital is injected.

Table 2.4 Gold–silver mine economic model

Year	UoM	Year 0	Year 1	Year 2
Production	tons ROM	–	60,000	60,000
Au grade	g/t	3.60	3.60	3.60
Au recovery	%	0.85	0.85	0.85
Au production	kg	–	184	184
Ag grade	g/t	210.00	210.00	210.00
Ag recovery	%	0.95	0.95	0.95
Ag production	kg	–	11,970	11,970
Revenue Au	$	–	74,58,750	74,58,750
Revenue Ag	$	–	63,59,063	63,59,063
Total revenue	$	–	1,38,17,813	1,38,17,813
CAPEX	$	1,35,15,000		
OPEX	$	–	76,80,000	76,80,000
Royalties	$		2,76,356	2,76,356
Cashflow before tax	$	−1,35,15,000	58,61,456	58,61,456
Tax (30%)	$		16,41,208	16,41,208
Cashflow after tax	$	−1,35,15,000	42,20,249	42,20,249

Source: The author

2.6.6 Project Financial Indicators

The project financial indicators are the key to an economic model. These are the indicators that any investor will look at to determine any further interest. There are three main indicators (Table 2.5).

An example of the project financial indicators for a gold project is summarised in Table 2.6.

2.6.7 Cash Flow Projections

The results of the economic evaluation of the mining investment opportunity can be presented graphically. Examples of a cash flow and cumulative cash flow for the gold–silver mine are in Figs. 2.18 and 2.19.

2.7 Valuation of Mining Projects

2.7.1 The Mine Developers' and Investors' Perspectives on Value

Valuation of mining projects is particularly relevant to equity investment. The mine developer and the investor have different entry points into a mining equity transaction:

Table 2.5 Project financial indicators description

NPV	Net present value A value used to determine the value of the project The net present value uses a discount rate which is applied to the cash flow In year 1 the cash flow is discounted by the discount rate, x In year 2 the cash flow is discounted by the discount rate, x2 In year 3 the cash flow is discounted by the discount rate, x3 High-risk mining projects have a discount rate of 20% Medium risk mining projects have a discount rate of 12% Normal mining projects are discounted by 10% The NPV is the sum of the annual cash flows, discounted
IRR	Internal rate of return This is a measure of the return of the investment Mining investors have hurdle rates from mining investment opportunities The minimum IRR for mining projects is 20% Investors to cater for inherent risks in mining The targeted IRR for mining projects is +25% from experience
Payback period	The payback period for mining investments is the period that is required to pay the investment (CAPEX) in the project Investors want the payback period to be as short as possible. Typical target payback periods are 3–5 years. An important consideration is the reserve tail. This considers the reminder of the life of the project after the CAPEX has been paid back. Investors want the reserve tail to be at least 30% of the project life to ensure that there is room to pay the capital back

Source: The author

Table 2.6 Gold–silver mine financial indicators values

Indicator	
NPV (10%) (US$)	1,12,87,818
IRR	29%
Payback period	4 years

Source: The author

- Mine developer: Secure funding to develop the project. Sell the minimum number of shares for the maximum amount of capital. However, do not scare the investor away.
- Mining investor: Acquire the maximum shares in mining project for lowest value. However, keep the mine developer motivated to pursue the development of the project.

Both the mine developer and the investor need to meet in the middle, should they be a willing buyer and seller, and both want to conclude the transaction.

This is where the independent valuation of the mining project, i.e. the investment opportunity, becomes important. It provides the required terms of reference for the negotiations between the mine developer and the investor.

Example: A tin mining project, which entails the reopening of an underground tin mine and the reworking of the tin tailings. The investor offers the mine developer an

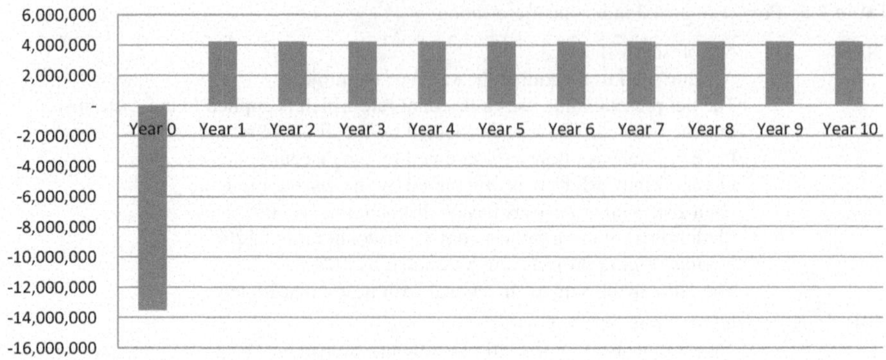

Fig. 2.18 Cash flow—gold–silver mine (Source: The author)

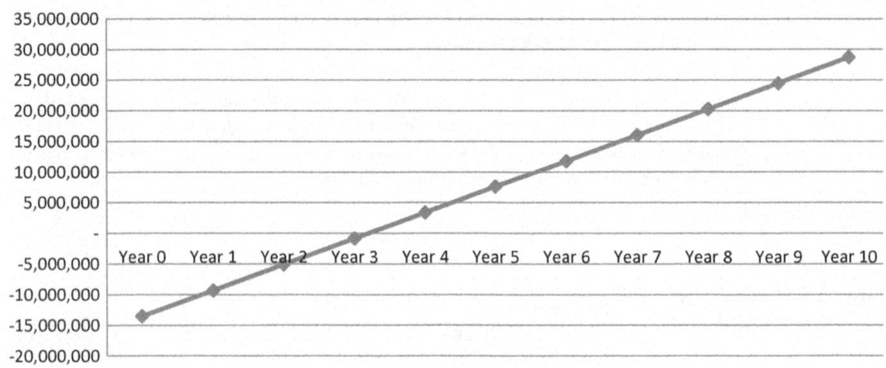

Fig. 2.19 Cumulative Cash flow—gold–silver mine (Source: The author)

acquisition fee of $3 million for 51% of the project, with the commitment to capitalise the entire project and pay a royalty to the mine developer. After lengthy negotiations, the mine developer declined the offer, hoping for a higher acquisition fee. Many years later, the mine developer is still looking for funding and the project has not started. Different perceptions of value led to this impasse.

2.7.2 The Increase in Value from Exploration to Production

The Mining Project Value Chain (Fig. 2.20) [44, 46] is a concept that is extensively used in mining investment and mine development circles as it encompasses various key aspects that are generic to all exploration and mining projects, regardless of which commodity is being pursued. It demonstrates how the value of a mining project increases from exploration to production.

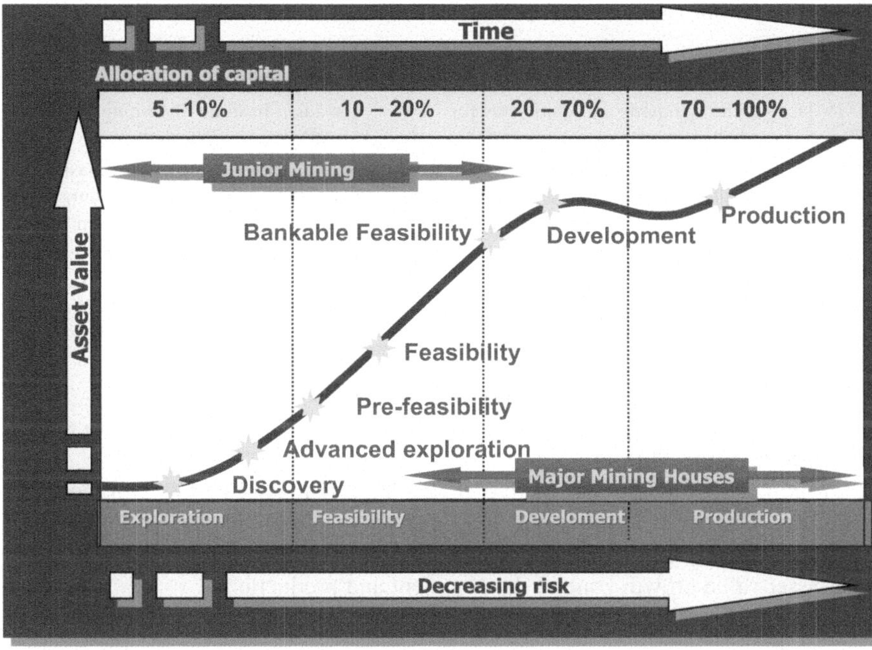

Fig. 2.20 Mining project value chain (Source: Wits CEE Course, Financing of mining projects, 2005)

_X_ Axis: Project Progression
The base line is the route along the value chain from exploration, feasibility, development to production.

_Y_ Axis: Asset Value
The asset value of the mining project increases rapidly from the exploration stage to the bankability stage. Thereafter, the asset value flattens, even decreasing, as the project now needs to meet the expectations of the investors. Only once production commences does the asset value curve grows.

Risk
The risk of the mining project is highest during the exploration stage and decreases as the project moves from exploration, feasibility, and development to production.

Capital Allocation
Only 20% of the capital is spent up to the bankable feasibility stage, as this is the high-risk zone. As the risk decreases, capital expenditure increases, with 80% of the capital being spent during the development and production stages. The bulk of the capital (50%) is spent during the development stage.

Table 2.7 Valuation codes and description

Valuation code	Description	Endorsed by
CIMVAL [47]	The standards and guidelines for valuation of mineral properties	Canadian Institute of Mining, Metallurgy and Petroleum
VALMIN [48]	The code for technical assessment and valuation of mineral and petroleum assets and securities for independent expert reports	Joint committee of the Australasian Institute of Mining and Metallurgy, Australian Institute of Geoscientists and the Mineral Industry Consultants Association
SAMVAL [49]	The South African Code for the reporting of mineral asset valuation	Southern African Institute of Mining and Metallurgy and the Geological Society of South Africa

Source: The author, based on CIMVAL, VALMIN and SAMVAL [48, 49, 50]

Junior and Major Mining

Junior mining companies are active in the high-risk zones of exploration and feasibility. This zone requires the least capital and presents the opportunity of highest asset value growth. Major miners play in the higher capital, lower risk and lower asset value growth zone of development and production.

2.7.3 Valuation Standards for Mining Projects

The development of mineral asset valuation standards is relatively new. These will assist mining deals being concluded, if both the mine developer and investor commit to using them. Three main international valuation standards for the mining and metals sectors have been established to date by the following professional institutions (Table 2.7):

2.7.4 Fair Market Value

The fair market value for mining project is the value which an independent mining expert has determined for purposes of concluding the investment transaction [5]. The value serves as a reference point for the investment between the willing buyer and a willing seller.

The fair market value is based on two considerations:

- The underlying technical value of the mining project as determined by the independent mining expert.
- A premium or discount related to factors such as the commodity market, host country or other considerations.

Table 2.8 Valuation methods and applicable project phase

#	Method	Applicable to project phase
1	Multiple of exploration expenditure	Exploration
2	Comparable transaction	Exploration, feasibility, development, to production
3	In situ metal value	Exploration
4	Discounted cash flow	Feasibility, development and production

Source: The author, based on mining industry principles

2.7.5 Valuation Methods

There are four main accepted valuation approaches used to determine a fair market value for a mining project. The valuation approach is dependent on the development status of the mining project [5, 47, 48, 49] (Table 2.8).

2.7.5.1 Multiple of Exploration Expenditure

This method is applied to exploration projects. The assumption is that exploration adds value to a property if positive results are obtained. If poor results are obtained, the value of the property decreases. A factor ranging between 0.5 and 3 is applied to the exploration expenditure.

Example: Gold exploration project:

Exploration expenditure: $1 million

The total value of the project ranges between $0.5 million and $3 million, pending the results of the exploration campaign (Table 2.9).

2.7.5.2 Comparable Market Transaction

This method is applied to all project development stages, from exploration, feasibility, development to production.

The assumption is that the value of the project is determined by current market transactions. The market, i.e. current transaction, takes the following issues into account in determining the $/t value for resources and reserves: stripping ratios, yields, the value of the mineral product, cost of mining and processing.

Example: Phosphate mining project.

The mine has 7.5 Mt of resources.

The total value of the project is $6.99 million (Table 2.10).

2.7.5.3 In Situ Metal Value

This method is applied to exploration projects.

Table 2.9 Multiple of exploration expenditure—gold exploration project

Results	Factor	Value of project
Very positive	3	$3 million
Positive	2	$2 million
Neutral	1	$1 million
Negative	0.5	$0.5 million

Source: The author, based on mining industry principles

Table 2.10 Comparable market transaction—phosphate mining project

Resource class	Mt	Value ($/t)	Value ($ million)
Measured	5.8	1.0	5.80
Indicated	1.7	0.7	1.19
Total value	7.5		6.99

Source: The author, based on mining industry principles

In early stage exploration, the economic value of the project is yet to be determined. A heavily discounted valuation of the total in situ metal contained within the resource is applied. This usually equates to a range of 2% to 4.5% of the spot commodity price as at the valuation date. Factors such as location, geological setting, infrastructure and the proximity to the market are considered.

Example: Gold exploration project:

In situ *metal: 1 million oz.*

Gold spot price: $*1100/oz.*

The total value of the project ranges between $22 *million and* $45 *million, pending location, geological setting, infrastructure and the proximity to the market* (Table 2.11).

2.7.5.4 Discounted Cash Flow

This method is applied to projects in the feasibility, development and operations stages. These projects have indicated and measured mineral resources. The prefeasibility study and feasibility study demonstrate a technically feasible and economically viable project, supported by calculated mining, processing, logistics, CAPEX and OPEX input parameters which were calculated by experienced mining professionals. The cash flow is discounted by an agreed discount value to determine the Net Present Value of the project. Applicable discount rates to be applied to the NPV calculations for the different project stages are outlined in Table 2.12.

Table 2.11 In situ metal value—gold exploration project

Results	In situ metal	Gold price	Factor	Value of project
Low value	1 million oz	$1100/oz	2%	$22 million
High value	1 million oz	$1100/oz	4.5%	$45 million

Source: The author, based on mining industry principles

Table 2.12 Project stage, risk rating and discounted rate

Project stage	Risk rating	Discount rate to be applied to Net Present Value (NPV) calculation (%)
Concept	High risk	15
Prefeasibility/ feasibility	Medium risk	12
Feasibility	Medium/low risk	10
Feasibility	Low risk	8

Source: The author, based on mining industry principles

Table 2.13 Discounted cash flow valuation—gold project

Results	Value of project
NPV	$4.5 million
IRR	26%

Source: The author

Example: Gold project

Project stage: Prefeasibility stage, some element of study at feasibility stage: medium risk

ROM: 1,500,000 t

Production rate: 120,000 tpa

Grade: 2 g/t

CAPEX: $13 million

OPEX: $16/t

Discount rate: 12% (Table 2.13)

2.8 The Importance of the Industry Cost Quartile Analysis

The position of the project OPEX on the industry cost curve 'Industry Cost Quartile Analysis' is an important indicator to determine competitiveness of a project in relation to its peers in the industry (Fig. 2.21).

The *X* axis depicts all the producers in an industry, e.g. the iron ore, platinum, coal or gold industry. The *Y* axis depicts the OPEX in $/t or $/oz.

The target for mine developers and operators is to have the OPEX of their mining business case in the lowest cost quartile, Quartile 1 or at least in Quartile 2.

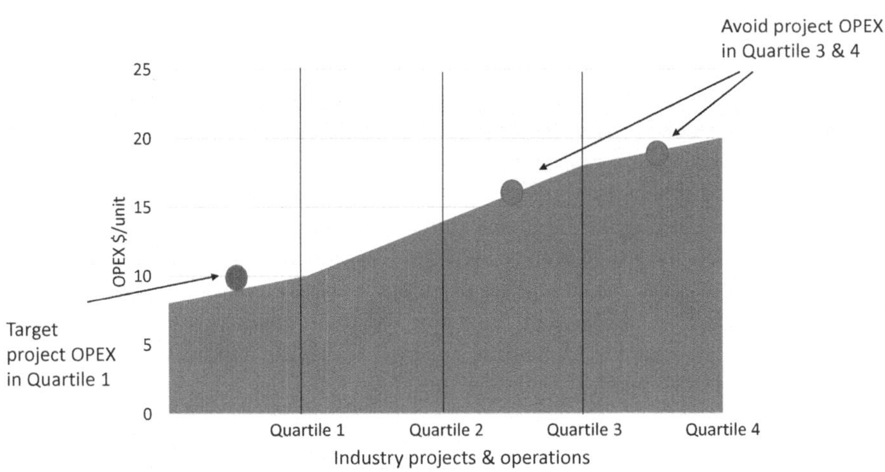

Fig. 2.21 Industry cost quartile analysis (Source: The author, based on mining industry practices)

During periods of strong commodity prices projects in all four quartiles are economic. However, when commodity prices decrease, projects in Quartile 4 are the first to be loss makers, followed by projects in Quartile 3 and then Quartile 2. Only projects positioned in Quartile 1 will remain economic.

Hence, investors should look at the position of the mining project seeking capital on the Industry Cost Quartile Analysis, and primarily invest in projects whose OPEX is in the first or second Quartile.

2.9 Risk Management for Mining Projects

2.9.1 Mining Projects Involve Risk

There are many mining projects that have collapsed with huge losses incurred for both the investors and mine developers.

Example:

A major mining company invested in an iron ore project in West Africa. A farm-in agreement was concluded with the local developers. Exploration commenced, the feasibility study was concluded, and construction works, and the purchase of large capital items commenced. Throughout these activities, the host government declined to convert the prospecting licence to a mining licence. The investor and mine developer challenged the host government to issue the mining licence but were not successful. The project collapsed with massive multimillion dollar financial losses incurred for both for the investor and developer.

This illustrates that there is a need for the early identification of risks and uncertainties before committing large sums to a capital-intensive mining project.

Mining projects have the following characteristics:

- Many new mining projects are developed in hostile geographical areas, where little experience of mining operations has been gained.
- New mining projects must compete with lower risk investment opportunities for capital. They need to offer higher returns as compensation.
- Mining projects have large capital outlays.
- There is a lead time before positive cash flows are generated.
- There is an increasing use of advanced technology.
- Mining projects must deal with stringent environmental and safety requirements.
- Mining projects after frequently exposed to new tax and royalty laws.
- Mining projects are dependent on many variables: geology, weather, market, technology, host country dynamics, finance, skills, etc. Some variables are in control of the developer, while some are not.

2.9.2 Risk Management Responsibilities

Responsibility of the Mine Developer

The mine developer needs to ensure that the investor gets substantial return on investment by managing the risks of mining project successfully. A mine developer who has a clear understanding of the project risks and has a risk management plan in place is more likely to persuade an investor to invest in the project, than a mine developer with limited comprehension of the risks or any risk management plan in place.

Responsibility of the Mining Investor

The mining investor needs to be fully aware of the risk profile of the investment that he or she is about to pursue and ensure that the mine developer has identified and quantified the risks and has a risk management plan in place. Most mining investors are aware that investment in mining projects is a high-risk undertaking, and hence a higher return on the investments is required. Mining investors must be willing to take a calculated risk when investing in mining projects.

2.9.3 Definition of Risk

Risk is a function of uncertainty (probability) and consequence (financial impact) [50].

Risk = Uncertainty (Probability) × Consequence (Impact).

The project developer needs to be able to quantify the probability and the impact of the risk.

The impact can be reduced production performance and revenues, and an increase in CAPEX and OPEX. The net result can range from lower than planned returns to a complete collapse of the project. The scope of risk assessment relevant to the project stage is outlined in Table 2.14. As mining capital is required during the feasibility and mine development stages, the mine developer needs to focus on risk assessment applicable to the first two project stages as a priority.

Table 2.14 Risk assessment applicable to project stage

Project stage	Scope of risk assessment
Feasibility	Identification of all technical, economic and human risks. Identify solutions.
Mine development	Risk assessment appropriate to source of funding. Identify solutions.
Operation	Monitoring of risks. Implement solutions.
Mine closure	Determine magnitude of risks that mining company should deal with when the project is closed. Implement solutions.

Source: Cooper, D. and Chapman C. Risk Analysis for large Projects—Models, Methods and Cases, Chichester and J. Wiley, Chichester, UK 1987

Table 2.15 Risks identified for mining projects

Risk category	Identified risks
Corporate risks	• Wrong strategic direction • Acquisitions—Wrong investment decision • JV partners/shareholders not aligned • Corporate governance
Operational risks	• Geology/exploration • Mining risks • Processing–poor yields • Infrastructure risks • Environmental risk • Weather—Force majeure • Production—Poor machine availabilities • Health and safety of employees compromised • Workforce—Industrial actions • Contractor non-performance • Completion risk • Project assumptions–incorrect • Escalating operating costs
Team risks	• Lack of skills • Lack of team cohesion • Lack of communication
Market risks	• Falling commodity price
Legal risks	• Mining regulatory framework inconsistent • Contractual risks • Country politics
Financial risks	• Financing—Funding terms, lack of finance • Currency movements • Insurance—Project not insured correctly • Participant credit risk

Source: Seeger, M. Development of a Strategic and Tactical Game Plan for Junior Mining Companies [44]

2.9.4 Risks Identified for Mining Projects

In Table 2.15, risks have been identified reviewing a series of mine development projects and categorised into corporate, operational, team, legal and financial risks [44].

2.9.5 Quantification of Mining Risks

Risk quantification is a process in which the probability of the risks identified, and the consequence thereof are workshopped in a team approach. Participants to this workshop include senior, experienced project members, whether in-house or external consultants of the disciplines: geology, mining, mineral processing, mechanical and electrical engineering, geotechnical, market, logistics, human resources, host country, environment and social, legal and finance (Fig. 2.22).

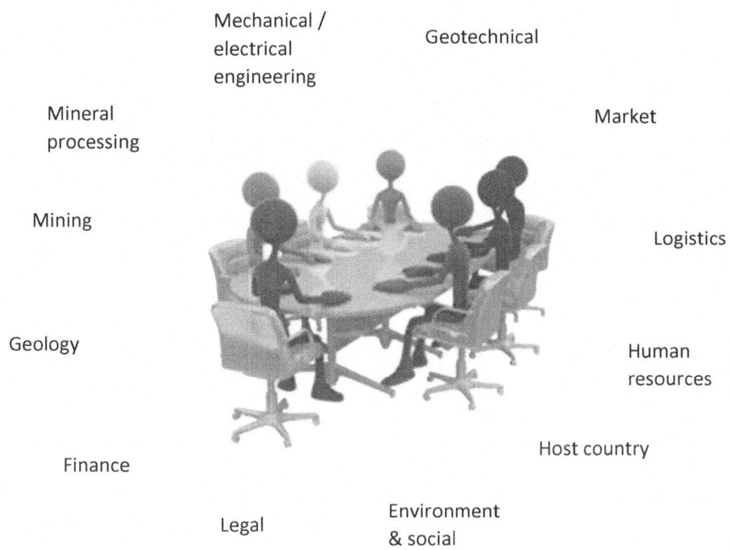

Fig. 2.22 Risk quantification for mining projects—workshop participants (Source: The author)

Participants in the workshop assign the *probability* and *impact* to each risk for the mining project:

Probability

Low probability: <20%
Medium probability: 20–50%
High probability: >50%

Impact
Expressed as a value, $ million. The financial damage incurred should the risk materialise.

An example of risk quantification exercise for a copper mining project is outlined in Table 2.16 presented below [51]:

2.9.6 Management of Mining Risks

The risk management plan needs to look at the high and medium risks. It is a summary that the developer needs to undertake to ensure the mining capital attracts the required returns, so the investor is satisfied. Below is a list of risk management initiatives that the reader can use as a reference (Table 2.17).

Table 2.16 Risk probability, impact and expected value—copper project

Risk	Probability low = <20% Medium: 20–50% High: >50%	Impact ($ million)	Risk expected value ($ million)
Production—Poor machine availabilities	40%	19.5	7.8
Escalating operating costs	38%	19.5	7.4
Contractors non-performance	35%	19.5	6.8
Project assumptions	40%	16.3	6.5
Country politics	25%	26.0	6.5
Currency movements	50%	13.0	6.5
Geology/exploration	30%	19.5	5.9
Falling commodity price	30%	19.5	5.9
Lack of skills	70%	6.5	4.6
Mining regulatory	70%	6.5	4.6
Financing	20%	19.5	3.9
Infrastructure risks	15%	19.5	2.9
Participant credit risk	15%	19.5	2.9
JV partners/shareholders	40%	6.5	2.6
Contractual risks	40%	6.5	2.6
Completion risk	35%	6.5	2.3
Mining risks	15%	13.0	2.0
Environmental risk	15%	13.0	2.0
Weather—Force majeure	10%	19.5	2.0
Insurance	5%	32.5	1.6
Workforce—Industrial actions	45%	3.3	1.5
Wrong strategic direction	20%	6.5	1.3
Processing—Poor yields	10%	9.8	1.0
Health and safety of employees	25%	3.3	0.8
Corporate governance	1%	13.0	0.1

Source: Seeger, M. Identification, quantification and simulation of risks facing greenfields mining projects in sub-Saharan Africa [51]

2.10 The Importance of Sustainability for Mining Capital

2.10.1 Why Sustainability Is Important to Mining Capital Providers

Environmental and social sustainability is a key consideration for mining capital providers, as their reputation and future business transactions are at stake, should the capital provider finance non-compliant projects. Mining capital providers place a high emphasis on mines being responsible corporate citizens with a strong

Table 2.17 Mining project risks, control and management

Risk	Control	Risk management
Geology/exploration	Control	• Thorough evaluation of geological information • Usage of internationally acceptable resource evaluation standards • Independent resource evaluation
Mining	Control	• Appointment of reputable mining consultants to provide monthly mine planning input • Use of reputable mining contractor • Experienced mine planners
Processing technology	Control	• Use proven technology • Input and guarantees from suppliers • Redesign if required • Recommendation from independent experts
Infrastructure risks–scarcity of water	No control	• Secure water supply upfront • Investigate alternative water supply scenarios: underground water, retreatment of water • Investigate processing methodologies with low water consumption
No infrastructure	No control	• Cannot influence
Weather—rain—force majeure	No control	• Due diligence to establish nature of such risks • Important to have consistency in force majeure clauses in all contracts • Insurance
Environmental risk	Control	• Implement environmental management report • Independent opinion • Ongoing monitoring
Production—poor machine availabilities	Control	• Focus on basics: water, oil, diesel, water, tools, cleaning of machines • Focus on preventative maintenance and regular services • Implement proactive maintenance management system • Search for and attract mechanical skills, incentivise • Focus on machine operator discipline and training to avoid abuse • Haul road maintenance • Engage OEMs and suppliers
Completion risk	Control	• Secure completion guarantees from sponsors • Ensure that sponsors have equity interest in completing project • Fixed price and fixed time turn-key contracts, guaranteed by reputable guarantor • Independent expert's reports before and after commencement of construction
Project assumptions	Control	• Review of assumptions by independent technical and financial advisors • Include a 10% contingency in inputs • Scenario planning
Lack of skills	Control	• Identify and promote talent in-house as far as practically possible • Partnerships with tertiary institutions to secure talent

(continued)

Table 2.17 (continued)

Risk	Control	Risk management
		• Create attractive working environment—reward performance • Attract 'old times' back into the company to share expertise and mentor young talent
Market forecast	No control	• Independent market study • Take or pay agreements • Price hedging arrangements
Mining regulatory	No control	• Ongoing consultation with the local department minerals
Country politics	No control	• Investment Protection Agreement • Political risk insurance • Presence of influential lenders such as World Bank Group, EIB • Foreign escrow accounts
Contractual risks	No control	• Legal due diligence process • Use well-tried legal jurisdiction
Financing	Control	• Fixed interest debt or bonds • Seek reputable financiers—Due diligence on lenders
Currency	No control	• Hedging • Matching construction and revenue currencies with loan currency • Offshore escrow accounts, if possible
Insurance	No control	• Diligent operational management • Insure equipment breakdown periods
Participant credit risk	No control	• Extensive credit evaluation of sponsors • Minimise the support required from weak sponsors • Costs of financing can be reduced where strong participants are prepared to give significant support

Source: The author

environmental and community stakeholder agenda. Any negligence of this important aspect in the development of a mine increases the risk of not achieving projected cash flows and reputational damage resulting in capital providers avoiding such projects.

2.10.2 Environmental Sustainability for Mining Projects

Many Examples of Poor Environmental Compliance by Miners

There are many examples of poor environmental compliance and environmental neglect by mining operations. Examples include mine acid drainage, polluted rivers near mines, mine wasteland, unrehabilitated mine voids and waste dumps, excessive mine dust, inadequate financial provision for rehabilitation.

On-Selling Mining Rehabilitation Responsibilities

A common practice is that the mines owned by larger companies are sold when the bulk of the reserves are mined out to smaller junior mining companies who have fewer financial resources. As the junior miner mines out the remnant reserves, at final closure there is often not enough capital available to rehabilitate the mine and the junior miner then files for voluntary liquidation.

Activist Environmentalists Can Cause Reputational and Financial Damage

If the community has a strong leadership, activist environmentalists and lawyers are engaged to sue the non-compliant mining company, its directors and shareholders. The damages claim can result in tens of millions of dollars which no mining financier will want to be associated with. This can ultimately lead to the collapse of the mining business.

Negative Image Scares Investors: Reputational Risk

Poor environmental compliance by mine developers and operators, and the associated negative impacts on community, water resources, fauna and flora has reputational consequences which negatively affect investment into mining. It creates a negative image for the mining industry and news spreads rapidly on social media platforms. It encourages critics of the mining industry to demonstrate against it, impacting the fundability of future projects, as prospective investors shy away from such reputational risks. To the investor, excessive environmental liabilities reduce the mine developer's ability to repay its loan, reduce the value of security and may raise the potential liability for the investor.

Changing Environmental Perceptions Affect Mine Financing Options for Coal

Changing environmental considerations and perceptions can affect the success or failure of financing of mining projects and the coal sector provides an example: Thermal coal mining is seen as a major contributor to global warming and many banks have decided to no longer provide project finance to coal projects to ensure that their reputation is not affected. This is forcing coal operations and mining projects to seek alternate financing options [7].

Rehabilitation Capital Provision

Rehabilitation commitment is an ongoing process throughout the operation of the mine. Usually a rehabilitation bond is registered with a finance house after the extent of mine rehabilitation at closure of the mine is calculated, considering removal of the surface infrastructure, contouring grassing of the mining pit, etc. An upfront deposit is paid prior to commencement of mining, and monthly instalments into the rehabilitation bond are made as soon as the mine is operational. A mining operation needs an approved environmental management plan and shareholders and management must commit to implementing it.

Equator Principles Required by Large Mine Finance Institutions

Many large financial institutions and Development Finance Institutions (DFIs) require mine developers applying for capital to comply with the risk management framework Equator Principles [52].

Equator Principles are used to determine, assess and manage environmental and social risks in mining projects and ensure a minimum standard for due diligence and monitoring. Mine developers are required to undertake an environmental and social impact assessment (ESIA), develop an environmental and social management report and undertake effective stakeholder engagement. The ESIA must comply with the host country legislation and apply the relevant IFC Performance Standards and the World Bank Group guidelines.

Mine developers need to comply with the following ten Equator Principles:

1. Review and categorisation
2. Environmental and social assessment
3. Applicable environmental and social standards
4. Environmental and social management system and action plan
5. Stakeholder engagement
6. Grievance mechanism
7. Independent review
8. Covenants
9. Independent monitoring and review
10. Reporting and transparency

Financial institutions applying Equator Principles carry out strict due diligence studies prior to entering into a loan agreement with mine developers and include protection such as representations, warranties, covenants and indemnities.

Not all mining capital providers require compliance with Equator Principles, as these can be very onerous on small junior miners.

Environmentally Complaint and Responsible Mine Developer: Attractive to Investors

The mine developer is forced to develop a business case that is environmentally compliant in order to attract the necessary mining capital. It is the right thing to do. The guiding principle is that the mining operation is a finite business. Once the ore has been mined out and the mining operations have been completed, the environment has to be returned to its natural state as best as possible. This means removing all structures of the mine, contouring the mine waste dumps, filling the mining pits with water, grassing. After rehabilitation is complete there is a process of ongoing monitoring. From the concept stage of developing a mining project, through to the pre-feasibility study, feasibility study to the mine design, environmental sustainability needs to be a key criterion. This will convince the mining capital provider that the company is a responsible corporate citizen, and that the risk of reputational damage is minimised, making it a more attractive investment destination.

2.10.3 Social Sustainability for Mining Projects

Minimising Downtime due to Community Unrest
Social sustainability plays an equally important role for mining capital providers. Mines have a finite life and work cooperatively with the community in order to minimise downtime due to strikes and unrest. Mining capital providers fear the financial consequences of loss of production, revenue, equipment if the mine is not a socially responsible citizen.

Raising and Not Meeting Community Expectations
Frequently, mine project developers promise benefits to the nearby community such as housing, a school, infrastructure, employment in order to secure the mining licence. Expectations are raised. Once the mining licence is secured, not all promises are held, and the community becomes negative and hostile to the mining project. When promises are not met and things go wrong, the community becomes hostile towards the mining project. In extreme cases, this may include barring of access to the mine, confiscating and burning mine equipment, and even taking the mine employees hostage. The root cause of all these actions is raised expectations that have not been met.

Local Employment Opportunities for New Mining Projects
A key consideration is local employment and the prospect of technical skills transfer to the local community. By offering young members of the community a career path and bursaries in technical and commercial mining related professions, community support is obtained. However, as mines get mechanised and automated, the opportunities for employment in large numbers become limited, and far less people will be employed by the mine as expected by the community. Again, the key is not to raise expectations too high.

Social and Labour Plan
Every mine needs a 'social licence' to operate sustainably. In some jurisdictions this process is formalised via a social and labour plan that must be approved which includes local economic development projects chosen from by the community. Each operating budget should have a $/t revenue provision to be allocated to a formal or informal social and labour plan.

Community Leadership and Interaction
An important consideration is the community leadership. Miners need to ensure that the leadership whom they are interacting with is the legitimately appointed community leadership, and not a fraction of the community. Community land and related surface right approvals become an important factor in the licensing process of mining projects. Delays in securing communal land delays the implementation of the mining project and affects the financing process, with mining capital providers moving on to other projects. Communication with the community on an ongoing basis is vital. Proactive mining companies deal with the social responsibility

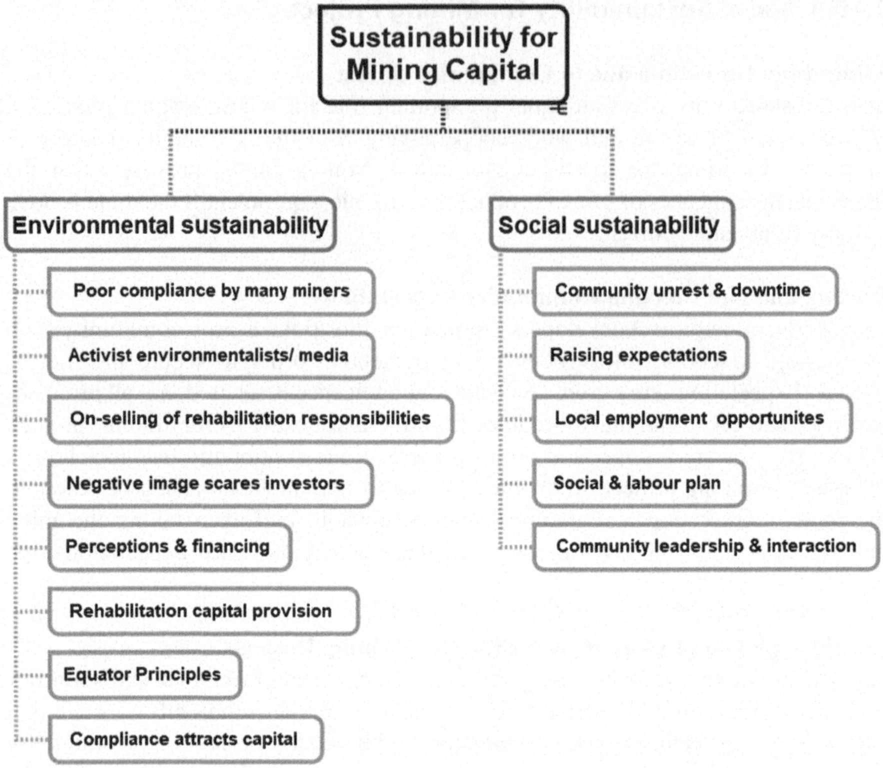

Fig. 2.23 Sustainability for mining projects—Key aspects for mining capital (Source: The author)

obligations by engaging the services of a local liaison officer. The scope of work of this position is to constantly engage and communicate with the affected community, facilitate local economic development projects and ensure positive community relationships.

The topic of environmental and social sustainability for mining projects is vast. This chapter summarised key aspects relevant to mining capital, depicted in Fig. 2.23.

2.11 Implementing the Mining Business Case: Project Plan Versus Reality

2.11.1 Project Plan: Base Case

Once exploration is complete, mining projects are implemented in a series of sequential project activities that follow each other: permitting, feasibility, finance, mine development and operations. The critical activities in the project implementation plan are the completion of the feasibility study, the issuance of the mining

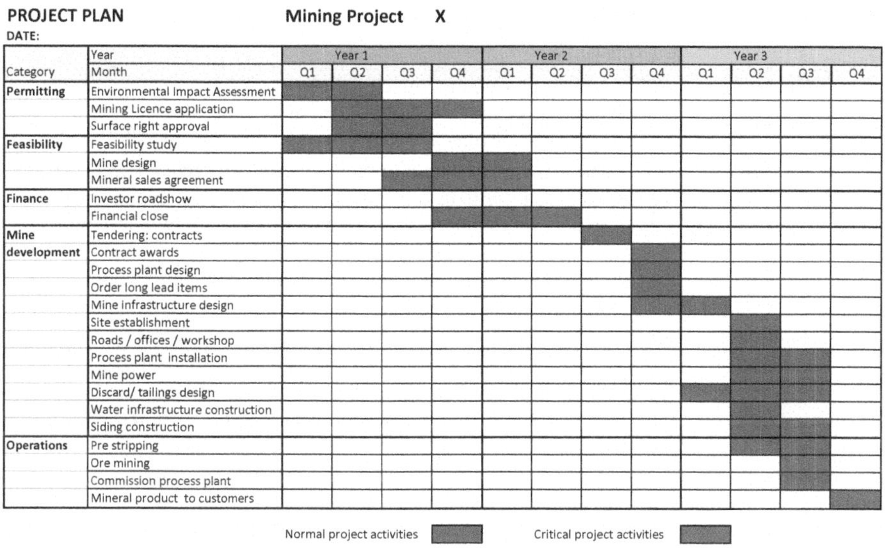

Fig. 2.24 Project plan—midsized mining project (Source: The author)

licence, the mineral sales agreement, the financial close, the award of contracts, site establishment, mining of ore and the mineral product to customer. Mining capital providers want to see a well-structured and thought through project plan. A typical project implementation plan for a midsized mining project having completed exploration is outlined in Fig. 2.24 and can serve as a reference. The activities remain the same for small, medium and large mining projects; it is only the timing that changes. Small projects can be implemented within 1–2 years, while larger projects take up to 10 years.

2.11.2 Project Plan: Reality

There are often delays in implementing mining projects and many mining deals take significantly longer than expected to be executed. Mine developers are often unrealistic in their expectations of project implementation timelines and mining capital providers are aware of overoptimistic timeline assumptions. Delays can occur due to various reasons and be short or long, pending the root cause of the delay (Table 2.18). Delays mean that the mining capital providers abort the financing process and the entire capitalisation process of financing must be restarted with new types of mining finance.

The example of a 2-year delay on the midsized mining project is outlined in Fig. 2.25. Following the delay, the feasibility study must be optimised, and the financing process is restarted. The project planned to be implemented within 3 years is only implemented in 7 years.

Table 2.18 Root causes for delay in mining project implementation

#	Category	Explanation
1	Shareholders	Shareholders change throughout the process of implementing the mining project and shareholder disputes for various reasons cause delays. Mining capital providers require shareholders to be aligned
2	Licencing	In certain mining jurisdictions the issuing of the Mining Licence takes significantly longer than anticipated. Change in governments, political interference, environmental considerations, community objections are one of the many reasons
3	Surface rights	Mine developers need to secure surface rights, and until a sale or lease agreement is concluded with the surface right holder, the mining project will not proceed
4	Commodity market	Commodity markets fluctuate and can change significantly during the process of developing the mine. This can render mining project uneconomic and the implementation plan is halted, until there is an upswing in commodity prices, failing which the project is mothballed
5	Mining legislation	Most host countries hold elections every 4 to 5 years. With the entry of new governments mining legislation is either continued or changed, pending social pressures and political dynamics of the host country. A change in mining legislation, impacting security of tender or royalties, may delay or halt projects
6	Financial market	Financial markets change constantly, and the nature of mining capital providers changes, impacting the timeline to financial closure. A recent example is several banks no longer investing in coal projects, causing coal project developers to seek mining capital elsewhere
7	Environmental	Project delays may occur as a result of negative environmental publicity caused by one mining project, affecting other similar mining projects, or changes in environmental legislation requiring the mining company to comply with further conditions
8	Team	Mining projects rely on a skilled and experienced mining team to implement projects successfully. A change of a mining team or the retiring of key skilled mining team members can leave a project in limbo until a skilled member is secured
9	Community	Communities have a say in the implementation of mining projects in their jurisdiction and may object to certain aspects of the project, with a resultant delay in implementation
10	Technology	Technological advances and disruptive technology can change the dynamics of mineral commodity supply and demand dynamics and may lead to mining project suddenly becoming uneconomic and having to be delayed or aborted

Source: The author

A representative sample of mining projects has been examined in terms of time distribution of mine development activities (Fig. 2.26). The trend is clear: The bulk of the time spent on developing mining projects is on feasibility studies and financing, while the development and construction activities are significantly shorter. This shows that the financing process, using diverse types of mining capital, is one of the longest processes in implementing projects and, mining project take significantly longer to be implemented than originally planned.

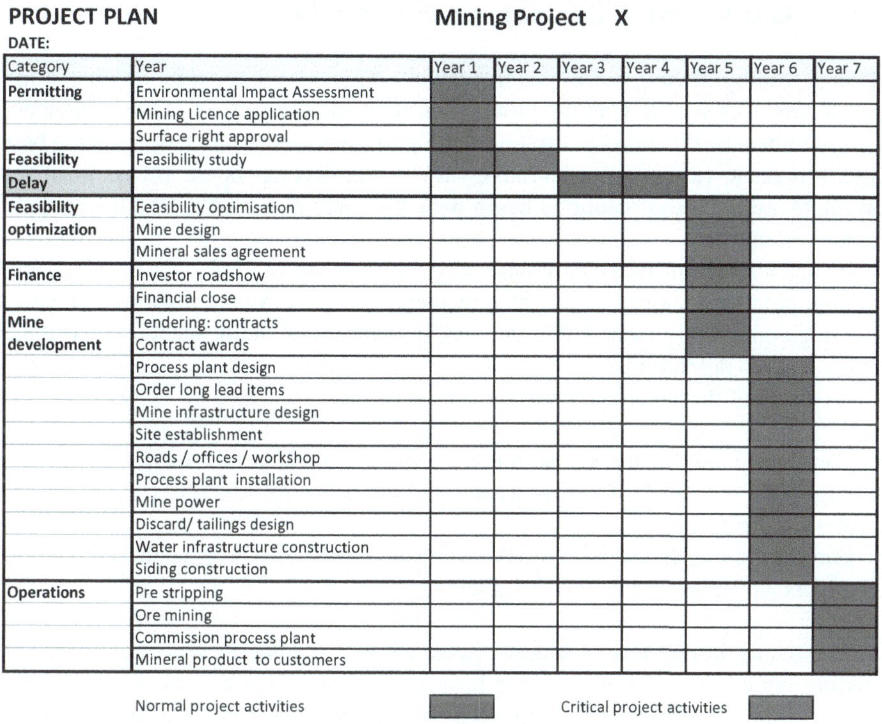

Fig. 2.25 Project plan including delay—midsized mining project (Source: The author)

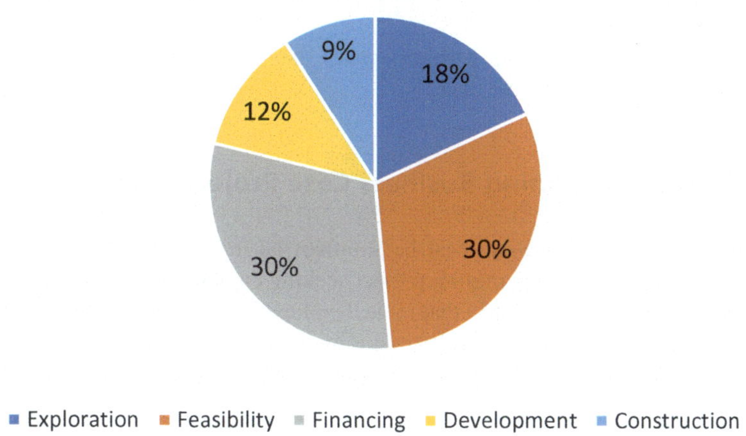

Fig. 2.26 Time distribution of mine project development activities (Source: The author)

2.12 Is the Mining Business Case Defendable

'Is the Mining Business Case Defendable?' This is a key question that mine developers need to ask. Can the production, revenue, CAPEX and OPEX numbers price numbers be defended in front of a group of investors? Is the business case best case, a realistic case or a conservative case? Mine developers are always positive; investors are often too sceptic to begin with and there is a need to find the realistic middle ground.

A defendable business case has supportive technical information (geology, mining, processing, logistics), legal agreements, licences, Memorandums of Understanding, quotations from EPC and O&M contractors, term sheets and offtake agreements to defend it.

As a rule of thumb and experience:

- Can the mining business case remain economic if only 75% of the production targets are achieved? If yes, proceed, but with diligence to the next level. If not, abort project.
- Can the mining business case remain economic if only 50% of the production targets are achieved? If yes, it is a winner.

The reality is that most business cases are full of imperfections and there are missing ingredients. If every mining business case were perfect, we would have a complete surplus of mines. The intentions of the guide are to get your business case as close to perfection as possible. The mine developer needs to understand the nature of these imperfections and still build the best business case possible with patience, tenacity, willpower. With a structured approach and a portion of luck, the investors will still invest in the project.

Details of imperfect mining business cases that cannot be financed and action plans cases are outlined in Sect. 4.9.

2.13 Summary: Mining Business Case Project Matrix

The Mining Business Case can best be summarised in a Project Matrix, in which 28 items are described for the mining project seeking capital. Table 2.19 provides the main headers of the Project Matrix that mine developers can use to compile the mining business case for later incorporation into a mining investor presentation and the mining business plan. It will also highlight missing parts of the mining business case.

Example: European Tin Project XXX

The Project Matrix has been completed using a European tin project as an example, entailing a mothballed underground tin mine and a tailings dam containing tin seeking $15 million to start Phase 1 of the project. The Project Matrix contains notes and guidelines in the right column highlighting key issues to be covered (Table 2.20).

Table 2.19 Project Matrix

Project name:
Commodity:

#	Item	Description
1	Project stage	
2	Location	
3	History	
4	Corporate structure	
5	Development team	
6	Permits and licences	
7	Geology	
8	Exploration	
9	Resources and reserves	
10	Mining	
11	Mine equipment	
12	Processing	
13	Production parameters	
14	Infrastructure	
15	Products	
16	Markets	
17	Logistics	
18	Human resources	
19	Environmental	
20	Society and community	
21	Financial—CAPEX	
22	Financial—OPEX	
23	Financial indicators	
24	Project plan	
25	Valuation	
26	Funding requirements	
27	Legal agreements/matters	
28	Documents available	

Source: The author

Table 2.20 Project Matrix—European tin project

Project name:		European tin project	
Commodity:		Tin (Sn)	
#	Item	Description	Notes/guidelines
1	Project stage	• Mothballed underground tin mine with tin-bearing tailings • Project stage: feasibility study • Phase 1: Tin tailings recovery project • Phase 2: Underground tin mining	• Exploration • Feasibility study • Development • Operations
2	Location	• European country xxx • Nearest town xxx, 20 km from the mine site • GPS coordinates: X0 Y′Z″ S, X0 Y′Z ″ E	• Country • Nearest town • Distance • GPS coordinates
3	History	• 1800s: Alluvial and underground tin mining • 1960–1970: Exploration by major tin mining company • 1980–2000: Underground mining by mining company • 2001: Mining company mothballed mine due to low tin prices	• Summary of historic activities to date • Companies involved
3	Corporate structure	• Holdco: European Tin Mining Company – Shareholder 1: 60% – Shareholder 2: 40% • SPV: European Tin Mining Company XXX	• Holdco • Shareholders • SPV
5	Development team	• Mine developer xxx, 15 years' experience in metals mining • Mining contracting company • Mineral processing company • Metals trading company	• Individuals • Companies • Experiences
6	Permits and licences	• Mining licence awarded for tin tailings recovery project and underground tin mine • Water abstraction permit for 5 million m3 annum • Surface lease agreement concluded	• Exploration licence • Mining licence • Water licence • Surface rights
7	Geology	• Tin veins hosted in the volcanogenic massive sulphide ore deposit	• Regional geology • Local geology
8	Exploration	• Historic underground mine data used to compile a new geological model and CPR • Aeromagnetic surveys completed • Exploration drilling: 55 holes, 3000 m • Tin tailings dump explored: 90 auger holes, 1200 m, 270 samples	• Geological surveys • Geological mapping • Soil sampling • Aeromagnetic surveys • Drilling (m/#holes) • Geological model

(continued)

Table 2.20 (continued)

Project name:		European tin project	
Commodity:		Tin (Sn)	
#	Item	Description	Notes/guidelines
9	Resources and reserves	Tin tailings: • Reserves: 2.5 Mt, 0.5% Sn Underground mine: • Indicated resources: 1 Mt, 1% Sn • Reserves: 1 Mt, 1% Sn • Competent person: xxxxx	• Reporting code • Inferred • Indicated • Measured • Competent person
10	Mining	Tin tailings recovery project: • Mining method: Hydraulic monitoring • Contractor mining and processing operation Underground mine: • Mining method: Vein mining, drill and blast • Contractor mining operation	• Mining method • Opencast/underground • Stripping ratio
11	Mine equipment	Tailings recovery project • 3 × hydraulic monitors (40 bar) • 3 × modular pump stations • 6 × 40 t ADTs • 2 × 75 t excavators • 1 × grader • 2 × D10 dozers • 1 × diesel bowzer • 1 × water cart • 1 × water cart • 5 × LDVs	• List of proposed mine equipment
12	Processing	• New modular tin processing plant to be established on site. Circuits include crushing, spiral plant, solvent extraction to produce tin concentrate 15% Sn • Capacity: 80 tph ROM • Tin concentrate to be smelted at smelter 20 km from site, to produce tin ingots, 100% Sn, via a toll smelt agreement	• Overview of mineral processing plant
13	Production parameters	Tailings recovery project: • Production: 480,000 tpa ROM (peak) • Head grade: 0.5% • Recovery: 50% • Tin production: 1200 tpa Underground mine: to be established	• Ore tons/month • Waste tons/month • Product tons/month
14	Infrastructure	Tailings recovery project: • Mine office • Power supply 5 MW • Mine security accesses	• Infrastructure requirement: tailings dam, power, water, access roads, buildings

(continued)

Table 2.20 (continued)

Project name:		European tin project	
Commodity:		Tin (Sn)	
#	Item	Description	Notes/guidelines
		• Processing plant • Tailings dam • Workshop	
15	Products	• Tin processing plant: Tin concentrate: 15% Sn • Tin smelter: tin ingots, 99% Sn, 1200 tpa	• Product specification • Quantity • Quality
16	Markets	• Global commodity trading company secured as offtaker for tin concentrate and tin ingots • Tin price: $15,000/t (99% Sn) • Payment based on %Sn in ingot	• Targeted customers • Pricing
17	Logistics	• Tin concentrate road transported 20 km to metals smelting company • Logistics cost: $2/t	• Description of logistics path: road, rail, ship • Distance to market • Logistics cost
18	Human resources	• Phase 1: Tailings recovery project to employ 60 employees • Phase 2: Underground tin mine to employ 150 employees—estimate	• Manpower requirements
19	Environmental	• EIA approved	• Status of EIA • Environmental permit
20	Society and community	• Mine employees to be employed from communities with 30 km radius from tin mine • Social and labour projects include bursaries in metallurgy for selected youth members of the community	• Community in project area • Community support for project
21	Financial—CAPEX	Phase 1: Tin tailings recovery project • Feasibility study and design: $1.0 M • Tin processing plant: $6.0 M • Tailings dam: $2.0 M • Mine infrastructure: $2.0 M • Rehabilitation bond: $1.0 M • Project management: $1.0 M • Contingency (15%): $1.8 M • *Total CAPEX: $14.8 M*	• Capital expenditure: mining, processing, infrastructure, logistics • Total capital expenditure
22	Financial—OPEX	Phase 1: Tin tailings recovery project $/ROM t • Mining: $2/t • Processing: $10/t • Logistics: $2/t • Social and labour plant: $1/t • Rehabilitation: $1/t • Management: $1/t • Total OPEX: $17/t	• Operating expenditure: • Mining $/t • Processing $/t • Logistics $/t • Total $/t

(continued)

Table 2.20 (continued)

Project name:		European tin project	
Commodity:		Tin (Sn)	
#	Item	Description	Notes/guidelines
23	Financial indicators	• NPV (10%): $7.8 M • IRR: 44% • Payback: 2.5 years	• NPV • IRR • Payback
24	Project plan	Phase 1: Tailings recovery project • Design: Q1, year 1 • Construction: Q3, year 1 • Commission: Q4, year 1 • Operation: Q1, year 2 Phase 2: Underground mine • Feasibility study: Q2, year 2 • Design: Q4, year 2 • Construction: Q1, year 3 • Commission: Q2, year 3 • Operation: Q2, year 4	• Timelines to production
25	Valuation	• Phase 1: Tin tailings recovery project: $7.8 M • Phase 2: Underground tin mine: $13 M	• Valuation of project • Expenses to date
26	Funding requirements	• $15 M for Phase 1 construction	• Total funding required ($) • Use of proceeds
27	Legal agreements/ matters	• Royalty agreement in place with original mining licence holder: 8% revenue-based royalty	• Agreements in place • Legal matters
28	Documents available	• Feasibility study—Tin tailings recovery project • Geological model underground mine • Quotation for processing contractor • Approved EIA • Mining Licence • Surface Lease Agreement • Metal Sales agreement draft	• Reports and studies • Business plan • Technical studies • Agreements • Financials • Other

Source: The author, based on a hypothetical tin project

Marketing the Mining Business Case: Best Practice

3

Contents

3.1 The Need for Marketing Material to Attract Mining Capital

The two key documents that market mining investment opportunity to potential investors are the mining investor presentation and the mining business plan. These documents need to be simple, concise and inform the prospective investor of the investment and return prospects. They are the key documents that are required to persuade the investors to invest their own capital into the mining business case. The key contents of both documents to be provided succinctly are location, ownership, licences, geology, mining, processing, infrastructure, team, market, logistics, financials, funding requirements, a risk management plan and the implementation plan (Fig. 3.1).

M. Seeger, *Mining Capital*, https://doi.org/10.1007/978-3-030-31225-1_3

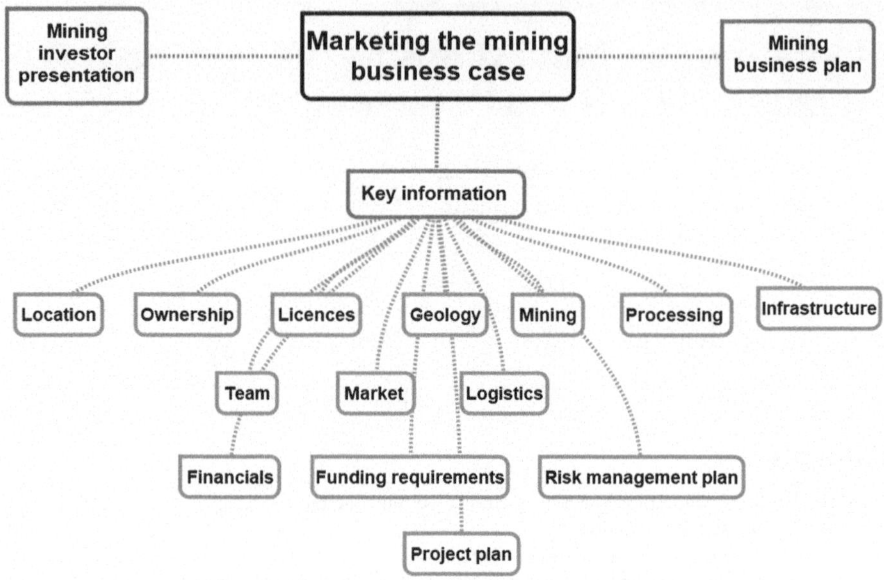

Fig. 3.1 Marketing the mining business case (Source: The author)

3.2 Mining Investor Presentation

3.2.1 Background to the Investor Presentation

The mining investment presentation is a key document and presentation that the mine developer needs to present to the targeted investors. This presentation on the investment proposition must be simple and concise, bring out the main points and be supported by graphics and pictures. The presentation must 'paint a picture' that attracts the investor. Many key investment decisions have been made primarily based on an investment presentation alone.

Each developer has his or her own preferences on how to present his/her investment opportunity. In my experience, it should be structured as follows:

- Project summary
- Location
- Ownership
- Licences
- Leadership
- Geology and exploration
- Resources and reserves
- Development strategy
- Mining
- Processing
- Mine infrastructure

- Logistics
- Market
- Development team
- Environmental and social
- Economic analysis
  - Input parameters
  - CAPEX and Funding requirements
  - OPEX
  - Economic indicators
- Project plan

3.2.2 Investor Presentation Example: ABC Gold–Silver Mining Project

A gold–silver mining project in South America, fully licensed and feasibility study completed, is seeking $13.5 million to develop the project. The project developers have spent $2 million to date on exploration, permitting and the feasibility study.

The presentation focusses on the project highlights. A phased development case is presented which generates cash quickly to expand the project at a later stage.

Fig. 3.2 Investor presentation Cover (Source: The author)

Project Summary

- Advanced gold project in xxx mining province, with significant silver credits
- Location: South American country xxx
- Fully permitted
- 3-phased development approach
- Phase 1: Development of small profitable underground gold and silver vein mining operation, which will produce 184 kg/annum gold and 11 970 kg/annum of silver
- Feasibility study completed & market secured for Phase 1
- Phase 2 & 3: exploration & feasibility to develop large scale opencast operation on tenement area
- 18 months to cashflow
- Capital required for Phase 1: $ 13.5 mill
- Phase 1 economics: IRR: 29%, NPV (10%) $11.3 mill, 4 year payback period

Fig. 3.3 Investor presentation—Project summary (Source: The author)

Location

- Host country: South American country xxxxx
- GPS coordinates: xxxxxx
- Nearest town: xxxx (100km from site)
- Site accessible via tar road xxx and 10 km of gravel road

Fig. 3.4 Investor presentation—Location (Source: The author)

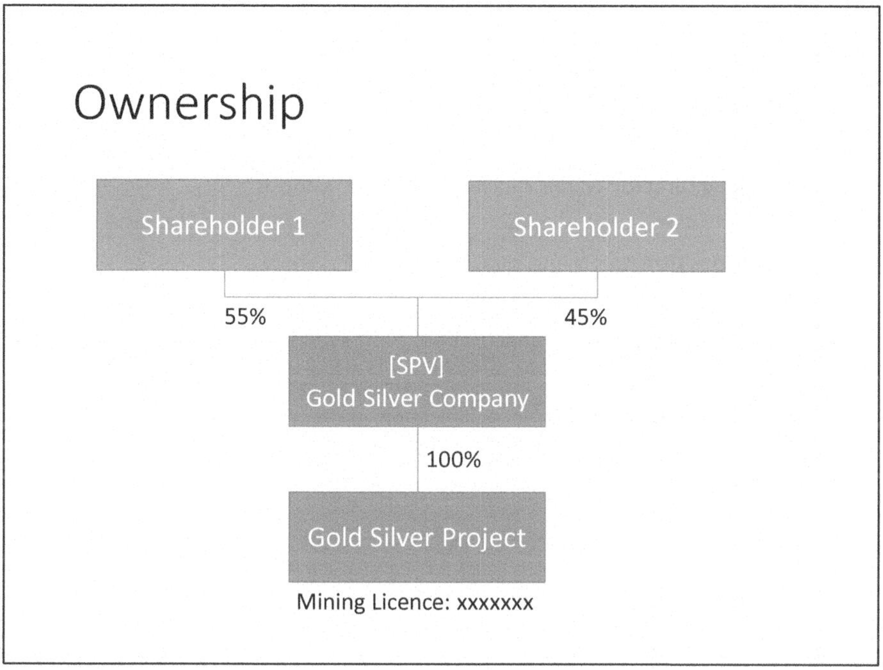

Fig. 3.5 Investor presentation—Ownership (Source: The author)

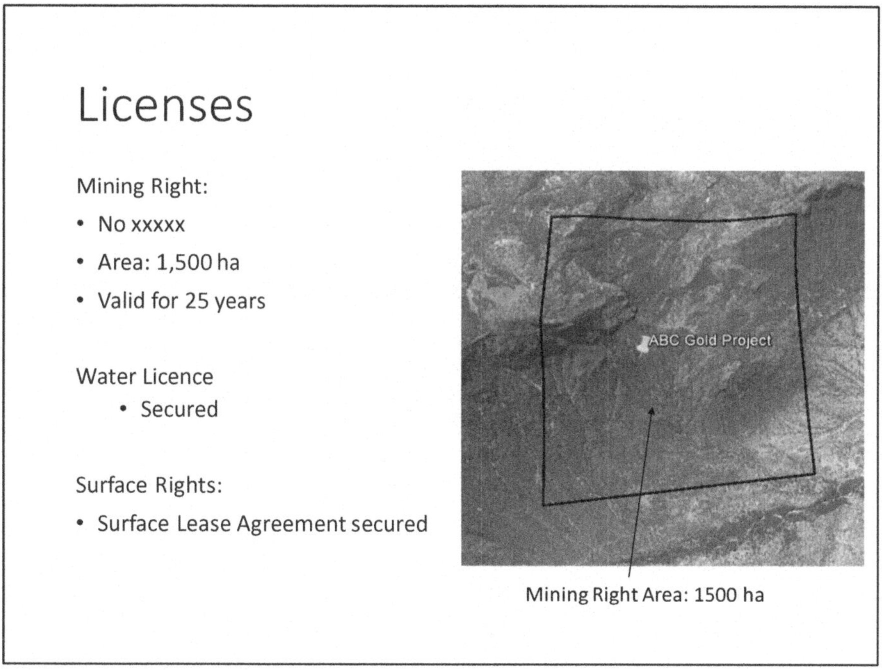

Fig. 3.6 Investor presentation—Licences (Source: The author)

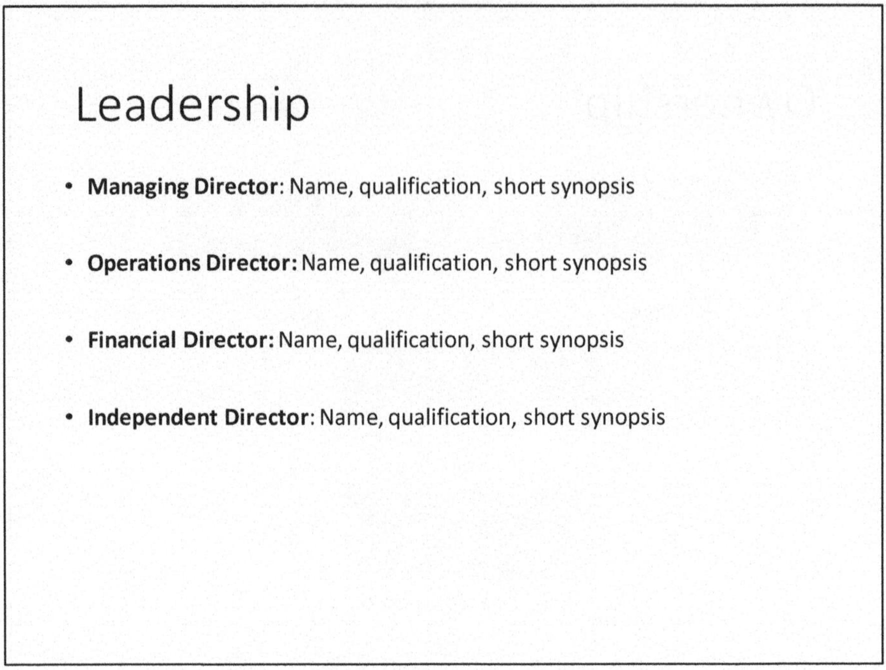

Fig. 3.7 Investor presentation—Leadership (Source: The author)

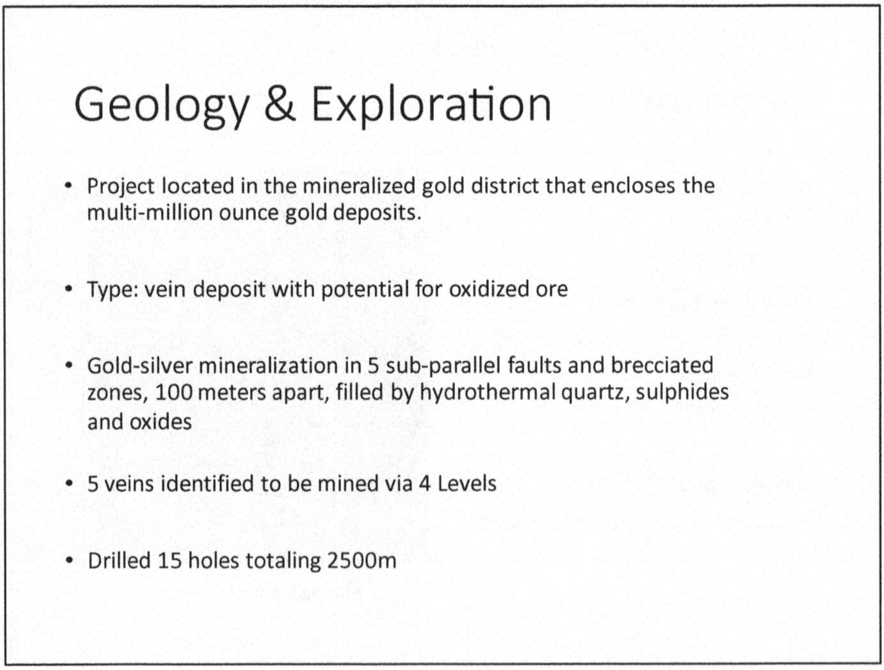

Fig. 3.8 Investor presentation—Geology and exploration (Source: The author)

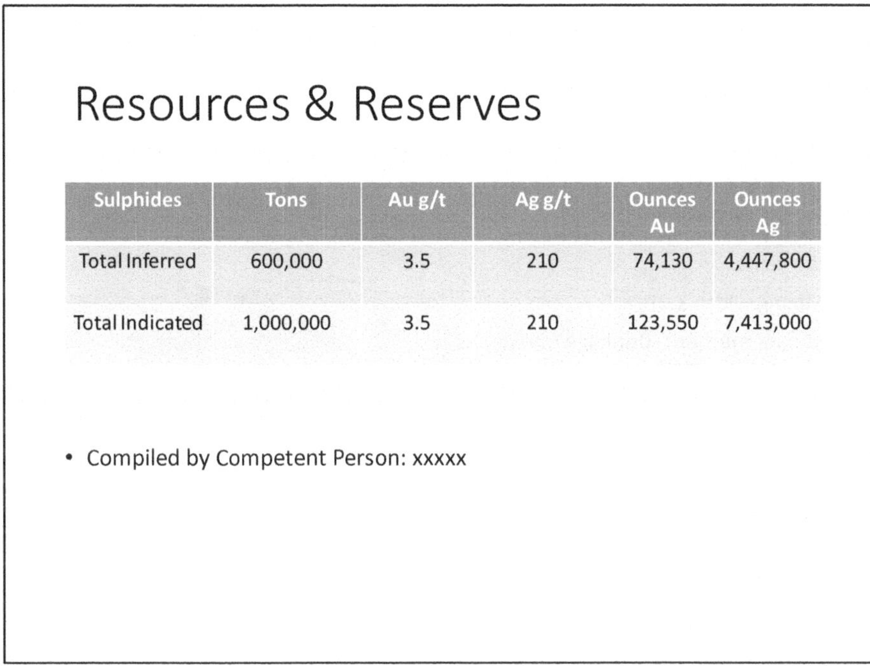

Resources & Reserves

Sulphides	Tons	Au g/t	Ag g/t	Ounces Au	Ounces Ag
Total Inferred	600,000	3.5	210	74,130	4,447,800
Total Indicated	1,000,000	3.5	210	123,550	7,413,000

- Compiled by Competent Person: xxxxx

Fig. 3.9 Investor presentation—Resources and reserves (Source: The author)

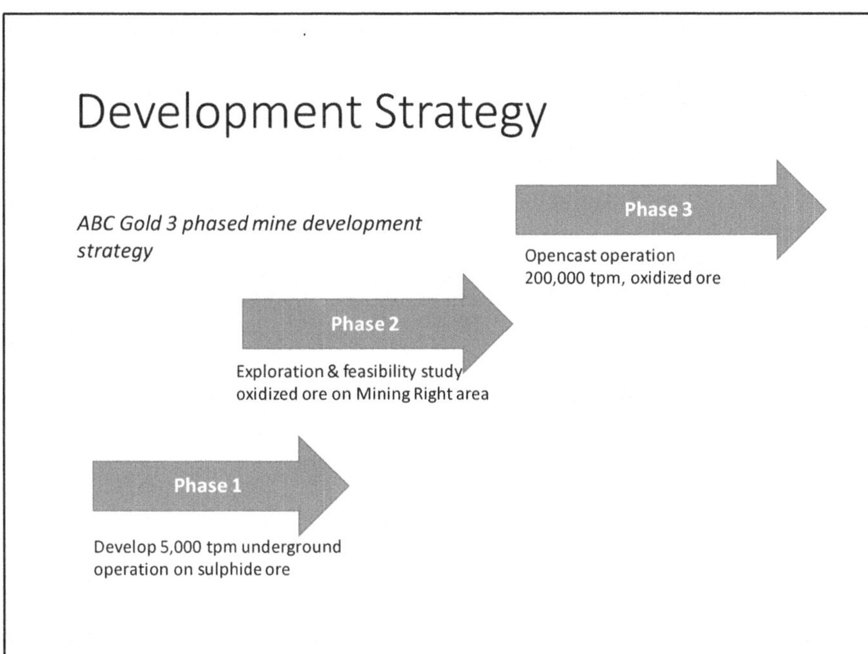

Fig. 3.10 Investor presentation—Development strategy (Source: The author)

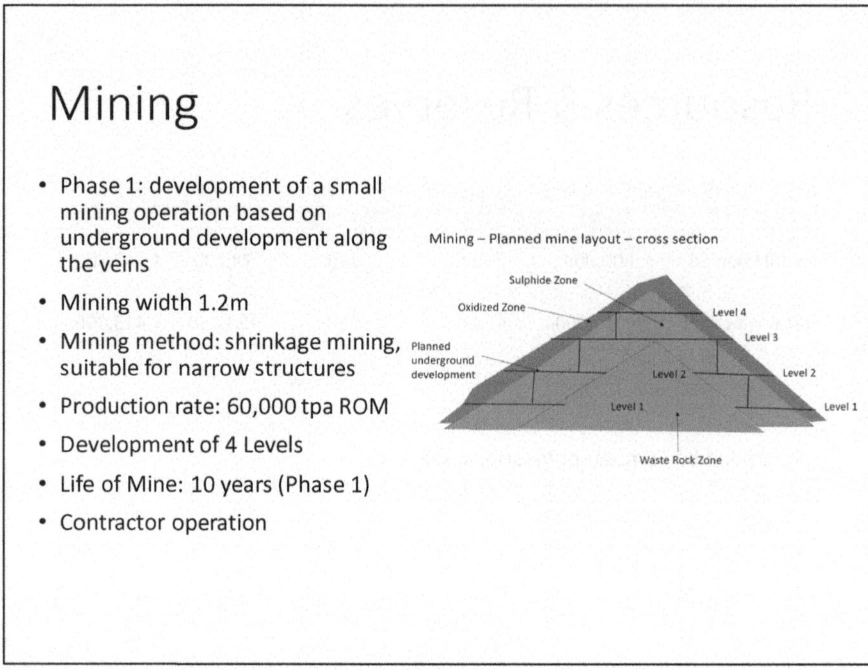

Fig. 3.11 Investor presentation—Mining (Source: The author)

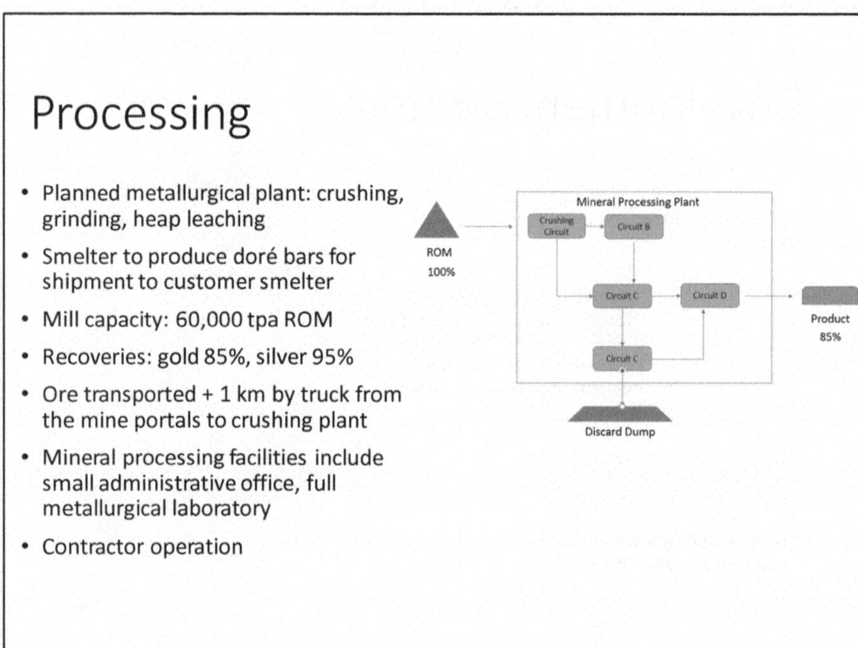

Fig. 3.12 Investor presentation—Mineral processing (Source: The author)

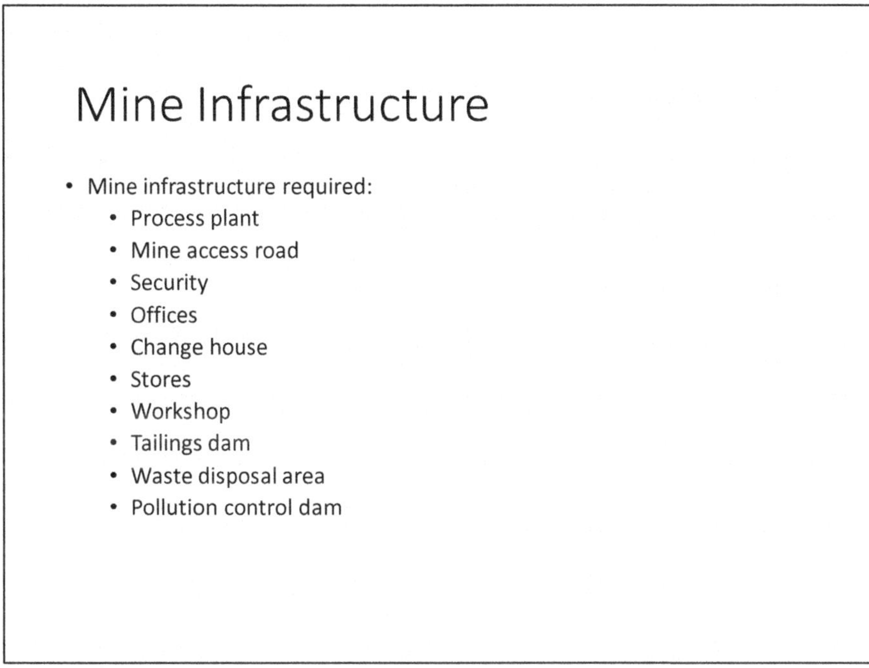

Fig. 3.13 Investor presentation—Mine infrastructure (Source: The author)

Fig. 3.14 Investor presentation—Market (Source: The author)

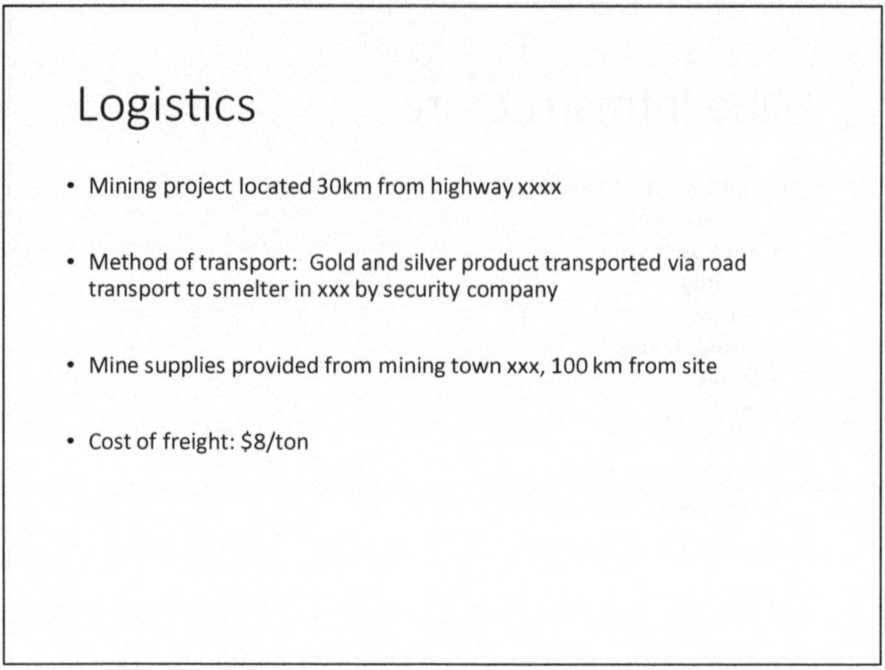

Fig. 3.15 Investor presentation—Logistics (Source: The author)

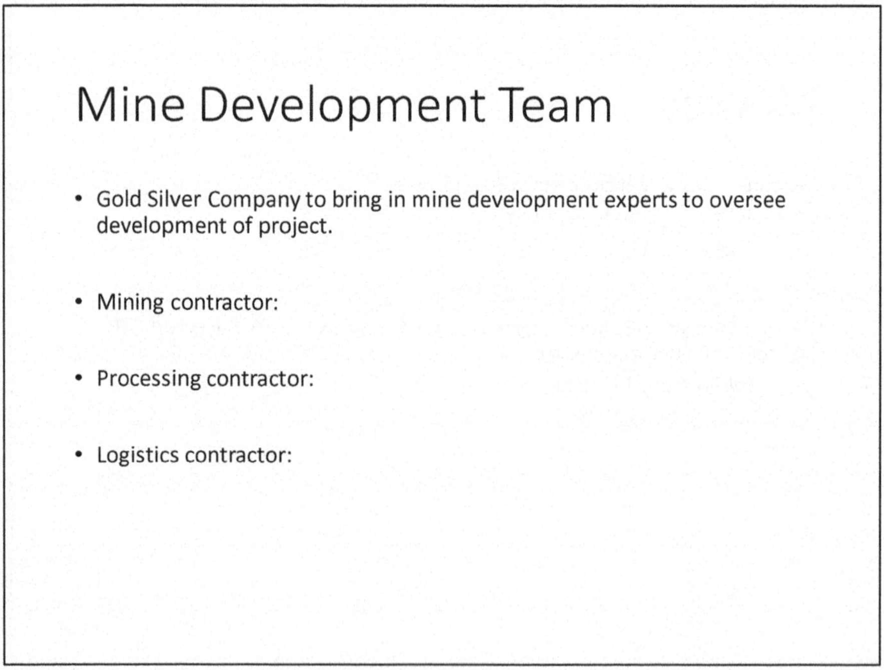

Fig. 3.16 Investor presentation—Mine development team (Source: The author)

Environment and Social

- Environmental Impact Study approved

- Planned community projects:
  - village school: materials and facilities improvement
  - health & development programme
  - community infrastructure

Fig. 3.17 Investor presentation—Environmental and social (Source: The author)

Economic Analysis

Input Parameters & Operating Costs

INPUT PARMETERS		
Parameter	UoM	#
Reserves	t	1,000,000
Production rate	tpa	60,000
Life of mine	Years	17
Au Grade	g/t	3.6
Au Recovery	%	85%
Au price	$/oz	1300
Au price	$/kg	40625
Ag Grade	g/t	210
Ag Recovery	%	95%
Ag price	$/oz	17
Ag price	$/kg	531.25
Tax rate	%	28%
Royalties	%	2%

OPEX	
Item	**$/t**
Mining	45.0
Processing	35.0
Logistics	8.0
Smelting & refining	26.0
Management	10.0
Environmental plan	2.0
Social plan	2.0
$/t	**128.0**

Fig. 3.18 Investor presentation—Economic analysis input parameters and OPEX (Source: The author)

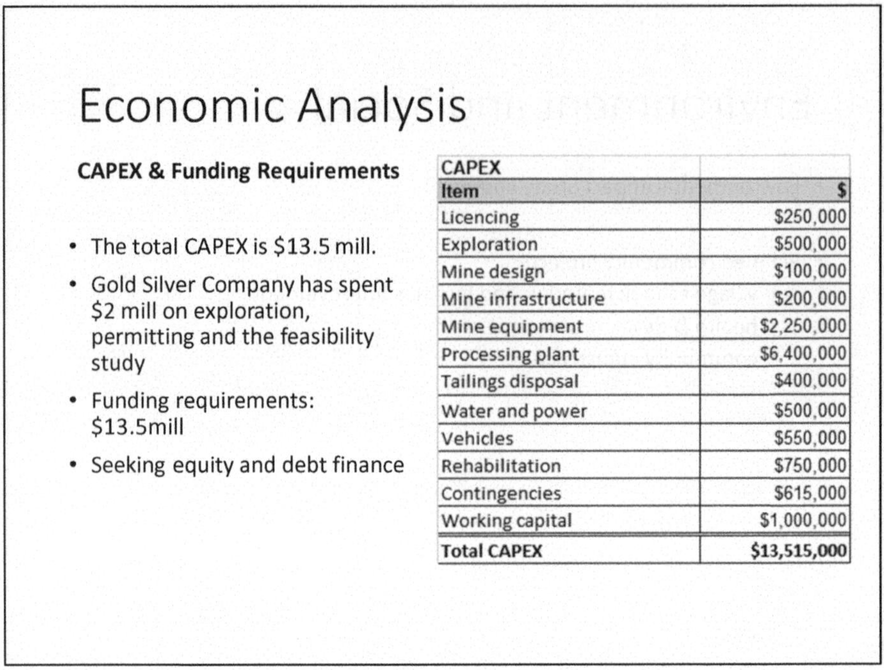

CAPEX	
Item	$
Licencing	$250,000
Exploration	$500,000
Mine design	$100,000
Mine infrastructure	$200,000
Mine equipment	$2,250,000
Processing plant	$6,400,000
Tailings disposal	$400,000
Water and power	$500,000
Vehicles	$550,000
Rehabilitation	$750,000
Contingencies	$615,000
Working capital	$1,000,000
Total CAPEX	**$13,515,000**

Fig. 3.19 Investor presentation—Economic analysis CAPEX and funding requirements (Source: The author)

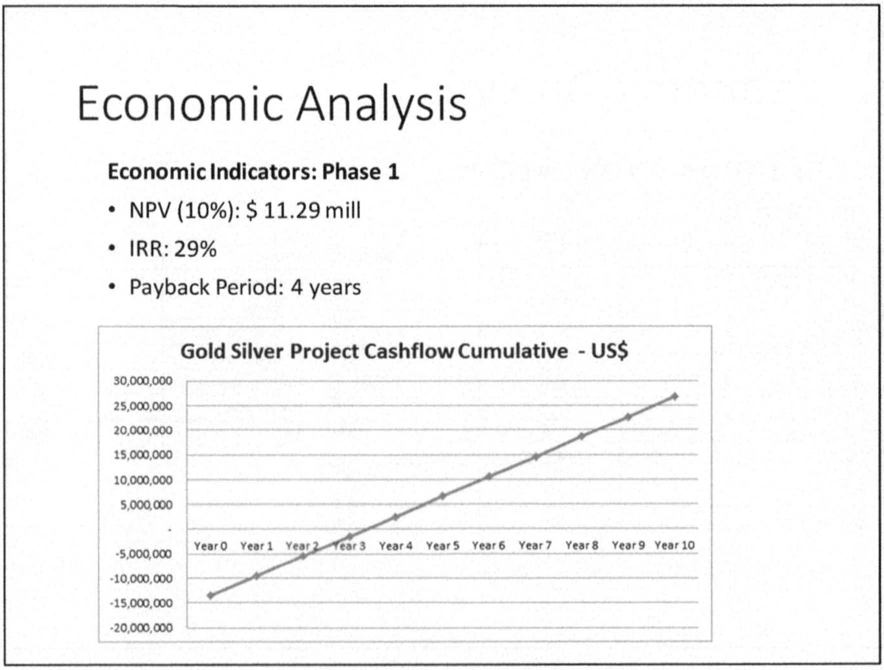

Fig. 3.20 Investor presentation—Economic analysis economic indicators (Source: The author)

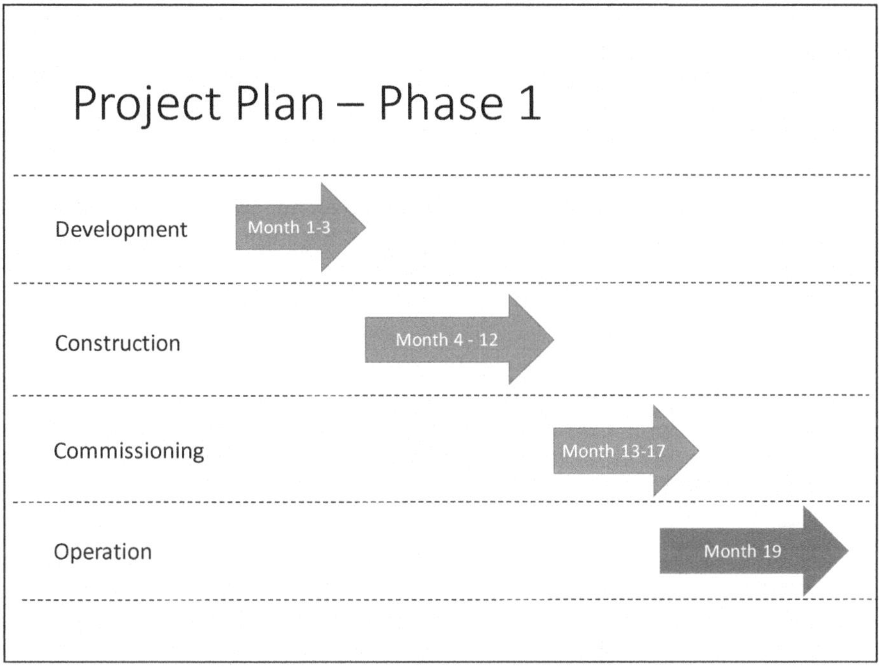

Fig. 3.21 Investor presentation—Project plan (Source: The author)

This investment presentation format can be modified to any commodity, stage of project or size of project (Figs. 3.2, 3.3, 3.4, 3.5, 3.6, 3.7, 3.8, 3.9, 3.10, 3.11, 3.12, 3.13, 3.14, 3.15, 3.16, 3.17, 3.18, 3.19, 3.20, and 3.21).

3.3 Mining Business Plan

3.3.1 The Key Document for Raising of Mining Capital

The mining business plan is the key document in securing funding for capital for mining projects and operations. It is a document that most institutions with mining capital require. Banks, development finance institutions, mining funds all require the mining business plan before considering investing into a project. Only traders may not need it if all required documents are in place.

Bottom line: The purpose of the business plan is to convince the targeted investors to invest in the business case.

No business plan = limited chance of money

The business plan is a 30–50-page document where all aspects of the business case are outlined and summarised in simple English. Note that decision makers often only look at the executive summary before choosing to continue reading or aborting. That means the executive summary is a key component of the business plan.

A strong business plan outlines in more detail what is presented in the investor presentation. In this chapter:

- A business plan structure is outlined that is recommended for mining business plans. This structure has been applied successfully in multiple mining finance campaigns.
- A successful business plan used for securing mining capital provided as an example. The mine developer reader can use this format for his or her own project.

Table 3.1 Mining business plan structure–Headers and comments

#	Header	Comment
1.	Executive Summary	Describe the investment opportunity, from geology, licensing, mining, processing, market to economics. The investor needs to be enticed to read on
2	Vision and Mission	
2.1	Vision	What is the vision of the project and its developer?
2.2	Mission	The project will produce × tpa of mineral product by date
3	Location	Country, nearest town, distance from market, GPX coordinates, accessibility
4	Ownership Structure and Shareholders	
4.1	Ownership Structure	The shareholder organogram
4.2	Shareholders	A short description of each shareholder. The investor wants to know who the potential partners are
5	History	When and by whom was the project initiated? When was it acquired by the current developer?
6	Licences	
6.1	Mining Licence	Mining licence number, for which commodity and duration
6.2	Environmental Licence	Status of environmental licence and EMP approval
6.3	Water Licence	The permission to extract water for mining operations
6.4	Surface Agreement	Who are the surface rights owners and what form of agreement has been struck?
7	Geology and Exploration	
7.1	Geology	Short description of the regional and local geology
7.2	Exploration	Outline what type of exploration has been conducted: sampling, geophysical surveys, aeromag surveys, drilling, trenching
8	Resources and Reserves	
8.1	Resource and Reserve Statement	A table showing measured, indicated and inferred resources and reserves
8.2	Competent Person	Statement of who the competent person is that signed off on resource reserve statement
9	Project Development Strategy	
9.1	Business Process	An overview of business process for the investor from geology, mining, processing, logistics to the customer
9.2	Implementation Strategy	Mining plans are usually implemented in phases. A short description of each phase

(continued)

Table 3.1 (continued)

#	Header	Comment
10	Mining	
10.1	Mining Method	A short description of the opencast or underground mining method
10.2	Mining Equipment	A summary of the mining equipment being used. Who will be capitalising the fleet?
10.3	Production Plan	The annual production profile of ore and waste
10.4	Life of Mine:	How many years of production?
11	Mineral Processing	
11.1	Mineral Process Plant 1	A short description of the mineral processing plant, including mineral circuits required to beneficiate the ore and produce the mineral products
11.2	Mineral Process Plant 2	As above
11.3	Mineral Process Plant n	As above
12	Mine Infrastructure	A table outlining the required mine civil, mechanical and electrical infrastructure for the project
13	Market	
13.1	Product and Prices	The mineral product produced, uses of the product, the buyer and offtake price secured
13.2	Customers	Description of the customer with whom an offtake agreement has/will be signed. Outline key terms of agreement
13.3	Competitors	A short description of the competitors and the mineral products that they produce
14	Logistics	
14.1	Distance to Market	How far is the mine located from the customer and key facilities
14.2	Mode of Transport	Road, rail, ship
14.3	Logistics Management	Who will manage logistics
15	Human Resources	
15.1	Organogram	A summary organogram of employees and contractors. It is important because it outlines to the investor the amount of jobs and people are affected by this investment
15.2	Mine Development Team	A short description of the team that implement the plan. Outline disciplines, names and experience. The investor will look at the credentials of the team when making the decision
16	Contracting	The key contracts and parties behind the business case
16.1	Contracting Structure	Display the investor the structure of legal agreements behind the business case in a simple flowsheet: mining, processing, infrastructure, and logistics contractors, customer, insurance, funder, technical and general services
16.2	Key Contracts and Suppliers	A summary of key contracts and suppliers
17	Environmental Management Plan	
17.1	Environmental Setting	Description of the environmental setting. Flat, mountains, trees, desert, bush, forest, arid, etc.
17.2	Environmental Management Plan (EMP)	A synopsis of key aspects of the EMP relevant to the project

(continued)

Table 3.1 (continued)

#	Header	Comment
17.3	Rehabilitation Cost	Provide the cost calculated to rehabilitate the mine after completion of operations. Is provision made in the operating costs?
18	Social and Labour Plan	An understanding of the community and sustainability to be displayed
18.1	Community	Describe the community in which the project is located. What is the level of employment? High unemployment levels mean there are high levels of expectation that need to be carefully managed
18.2	Local Economic Development	List the initiatives that the mining company will pursue and commit to in to developing the local community
19	Financial Analysis	
19.1	Input Parameters	These are the input parameters into the economic model: ROM ore (Mt), production ore (tpm), stripping ratio, production waste (tpm), Life of Mine, ore grade, mineral processing recovery %, product prices, royalties, tax rate
19.2	CAPEX Requirements	A summary list of the CAPEX ($mill) to develop the project/business case: mining, processing, mine infrastructure, power and water, project management and a 10% contingency
19.3	Operating Costs	A summary list of operating costs ($/ton sales) for mining, processing, transportation, management and rehabilitation
19.4	Project Cash Flow	A graphical depiction of the cash flow for 10 years or the Life of Mine
19.5	Financial Indicators	The three important indicators are: Net Present Value (NPV) at a discount rate, Internal rate of return, Payback period
19.6	Funding Requirements	A statement of what funding is required from the investor and what that debt equity ratio is
19.7	Expenditure to Date	How much has been spent on the development of the project to date? The investor will see this as equity contribution in determining future shareholding percentages
20	Risk Management Plan	
20.1	Technical Risks	A short summary of the relevant geological, mining, processing, infrastructure and production risks, their ranking (high, medium, low) and the risk management plan
20.2	Human Risks	A short summary of the relevant community, political, host country risks, their ranking (high, medium, low) and the risk management plan
20.3	Economic Risks	A short summary of the relevant market, financial, and contractual risks, their ranking (high, medium, low) and the risk management plan
21	Implementation Plan	A high-level Gantt bar chart of the project implementation plan
22	Conclusion	*Summarise the key considerations why the mining business case is an attractive investment proposition, supported by key economic indicators*

Source: The author

3.3.2 Mining Business Plan Structure

The structure of a business plan for mining projects, irrespective of commodity, project stage and location, is outlined in Table 3.1.

3.3.3 Mining Business Plan Example: Vanadium Titanium Iron Project

A winning mining business plan example is provided in this chapter, that was used to raise CAPEX for a vanadium and metals project. For confidentiality purposes, any names or references have been deleted and are referred to as 'xxx'. This mining business plan can be amended to any mining business case within any commodity space.

Vanadium Titanium Iron Project: Business Plan

1. Executive Summary

The Vanadium Titanium Iron (VTI) Project is a mine development project owned by VTI Resources in xxx that entails the development of a titaniferous, vanadiferous magnetite deposit with a mining licence, market and a completed feasibility study. The project forms part of VTI Resources mission to become a significant vanadium, titanium and pig iron producer by 20xx.

The project is strategically located with direct access to the main export port of xxx, via rail. The project has been extensively explored since the 1980s, and several exploration campaigns have been conducted, totaling 10,000 m and 2000 assays. VTI Resources has secured all the necessary permits to begin with mining operations.

The project is based on a single 20 m thick titaniferous magnetite seam that contains vanadium (V_2O_5), titanium (TiO_2) and iron (Fe). The seam extends over an 8 km strike length. Weathering has occurred on surface, where a titanium bearing sand deposit has formed. The VTI Project has 50 Mt of measured and inferred resources of grading 18% TiO_2, 35% Fe and 1% V_2O_5.

The project entails an opencast mining operation, a mineral concentrator and a smelter to produce vanadium slag, titanium slag and pig iron. Mining operations will produce 0.6 Mtpa of ROM for the mineral concentrator and will be conducted by a reputable mining contractor. The mineral concentrator will produce 0.24 Mtpa of heavy mineral concentrate (45% TiO_2). The heavy mineral concentrate will be transported via road and rail to a smelter, where the concentrate will be smelted to produce titanium slag (85% TiO_2), pig iron and vanadium slag. VTI Resources has concluded a 10-year toll smelting agreement with an existing smelter which eliminates the need to build a new smelter.

VTI Resources has further concluded an offtake agreement with metals trading company xxx, which will purchase all the titanium slag, vanadium slag and pig iron for the first 5 years of production at an agreed offtake price. This will significantly lower the risk of the VTI Project, as the market is secured for 5 years.

The VTI Project will employ 200 employees, the majority of whom will be employees of the appointed mining and processing contractors. An expert mine development team has been appointed by VTI Resources into production.

Key contracts have been concluded with the mining contractor, processing contractor, logistics, metals trading company and service providers.

The Environmental Management Plan has been approved by the authorities and VTI Resources has been lodged the rehabilitation bond. The local economic development initiative is centred around the development of technical skills to create opportunities for the community.

The project's economic parameters are based on mining, processing and smelting of 5.7 Mt of magnetites ore over 10 years, at a rate of 50,000 tpm to produce 1,21,500 tpa of titanium slag, 81,000 tpa of pig iron and 1080 tpa of vanadium slag. The CAPEX requirement is $34 million and the OPEX is $341/ton. The project has a NPV of $35 million at a discount rate of 10%, an IRR of 31% and the payback period is 4 years.

The shareholders of VTI Resources have spent to date $2.5 million on the acquisition and development of project. The CAPEX requirement of $34 million and has a gearing ratio of 60% debt and 40% equity. VTI Resources is exploring various types of funding to bring the VTI Project into production including contractor financing, development finance, offtake finance and equity finance.

2. Vision and Mission

2.1. Vision

The vision of VTI Resources is to establish a sustainable, profitable high-grade vanadium titanium iron mining project by 20xx, returning maximum benefit to all stakeholders. The project presents an ideal investment opportunity for investors in a mine with strategic future commodities.

2.2. Mission

By 20xx, VTI Resources will produce 1,21,500 tpa of titanium slag, 81,000 tpa of pig iron and 1080 tpa of vanadium slag from the VTI Project.

3. Location

The VTI Project is located in xxx, 150 km from a main minerals port. The nearest town is xxx, 5 km from the site.

VTI Mine is accessible via the main tar road xxx and a 10 km gravel road.
GPS coordinates: X° Y′Z″ S, X° Y′Z″ E

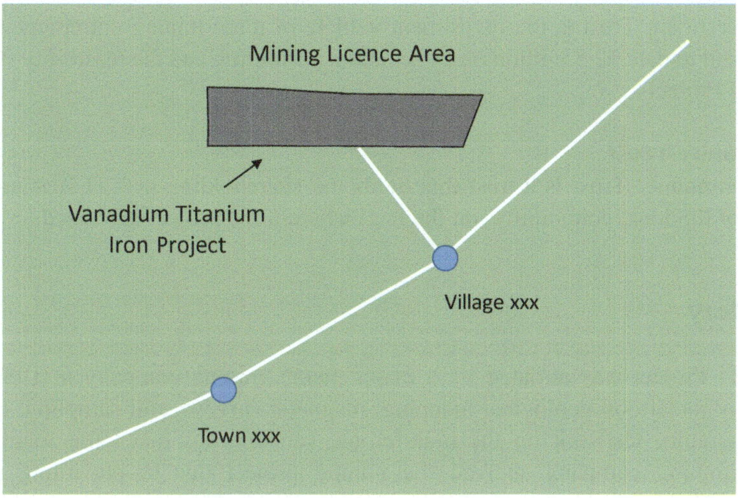

Source: The author

4. Ownership Structure and Shareholders

4.1. Ownership Structure

The Vanadium Titanium Iron (VTI) Project has the following ownership structure:

Vanadium Titanium Iron (VTI) Resources the 100% shareholder of Vanadium Titanium Iron Project (Project SPV).

The shareholders of VTI Resources are the investment company (85%) and community trust (15%).

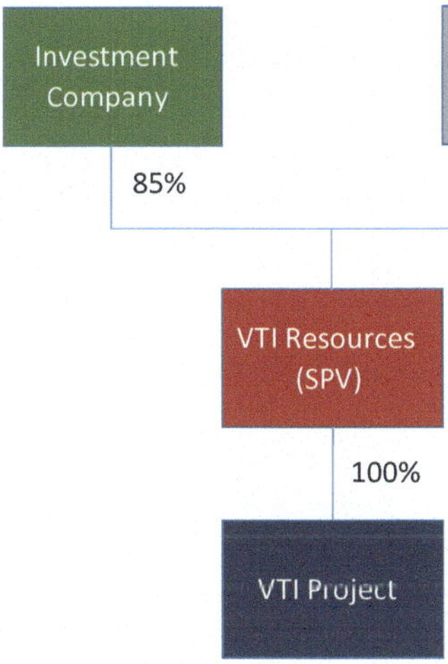

Source: The author

4.2. Shareholders of VTI Resources

Investment Company
The Investment Trust is the investment vehicle of three mining entrepreneurs that have spent to date $2.5 million on exploration, permitting and the feasibility study of the VTI Project.

Community Trust
The Community Trust is a trust that holds the shareholding in VTI Resources on behalf of the local community, on the land wherein the project is located.

5. History

The VTI Project was initiated by a major metals mining company in 1980. The company undertook geological mapping, magnetic surveys, soil sampling, drilling and laboratory testwork on the titaniferous, vanadiferous magnetite seams. The metals mining company shelved the mining project due to prevailing market conditions.

VTI Resources acquired the VTI Project in 20xx and commenced with confirmation drilling and the feasibility study.

6. Licences

VTI Resources has secured all the necessary permits to conduct mining operations. The relevant State Departments are all keen to see the development of the VTI Project.

6.1. Mining Licence
VTI Resources has been granted Mining Licence No. xxxx. The licence grants VTI Resources the exclusive right to prospect and mine the vanadium-rich magnetite seams of a 25 km^2 area. The mining licence is attached.

6.2. Environmental Licence
VTI Resources was granted the environmental licence after approval of the Environmental Management Report.

6.3. Water Licence
VTI Resources has secured Water Licence xxx to extract 3,00,000 m^3 of water per annum from the xxx river.

6.4. Surface Agreement
VTI Resources has an agreement with the community, the owner of the property, to conduct mining operations on the tenement area in perpetuity for a royalty.

7. Geology and Exploration

7.1. Geology

The vanadium-rich magnetite seams of the VTI Project are hosted within the xxx complex. The seam strikes at 80° south at an average dip of 85°.

The 20 m thick vanadium-rich magnetite seams contains vanadium (V_2O_5), titanium (TiO_2) and iron (Fe).

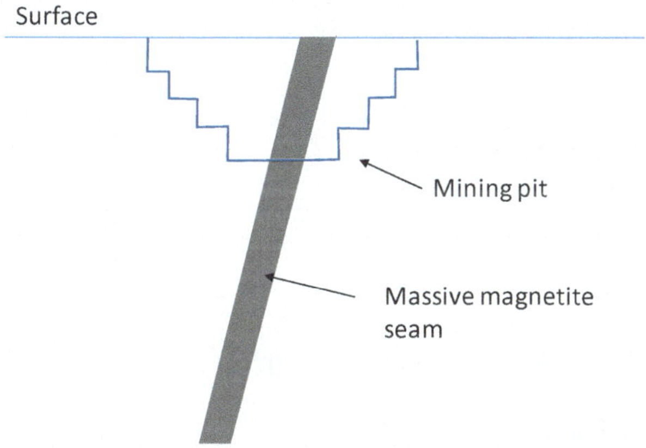

Vanadium-rich magnetite seam outcrop (Source: The author)

7.2. Exploration

The following exploration has been conducted on the VTI project:

Geological Mapping and Field Reconnaissance
Ground geophysical survey:

- A high-resolution ground magnetic survey was completed on the property.
- The magnetic survey results outlined the outcrop and sub-outcrop of the orebody and the occurrence of massive magnetite seam.

Drilling:

- Total m drilled: 10,000 m
- Type of drilling: diamond drilling
- Total assays: 2000

Trench sampling:

- 20 trenches, 1.5 m deep, were dug and sampled.

Geological modelling:

- A 3D geological model created from the drilling information.

8. Resources and Reserves

8.1. Resource and Reserve Statement

Resource	Mt	V_2O_5 (%)	TiO_2 (%)	Fe (%)
Measured and Indicated	30	1	18	35
Inferred	20	1	18	35
Total	50	1	18	35

Source: The author

8.2. Competent Person

The competent person that compiled this statement is xxxx, an experienced and well-respected vanadium geologist.

9. Project Development Strategy

9.1. Business Process

The business process from mining to product is outlined below:

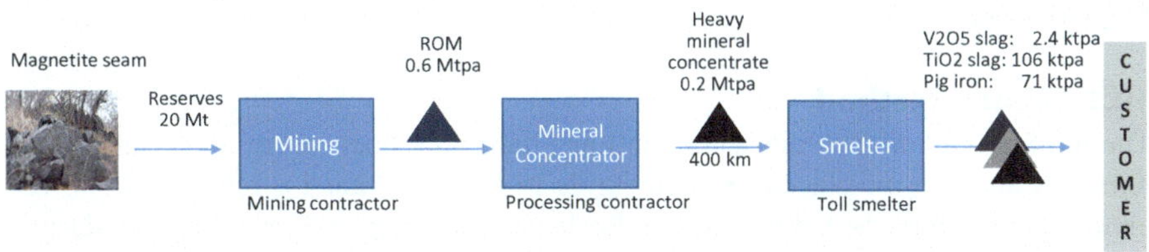

Source: The author

Mining operations will entail opencast drill and blast operations at a 3:1 stripping ratio, conducted by a mining contractor, to produce 0.6 Mtpa of vanadium-rich magnetite ore.

The ore will be fed into a mineral concentrator. The mineral concentrator entails crushing and milling circuits, a spiral plant and magnetic separation circuits to produce 0.24 Mtpa of heavy mineral concentrate.

The heavy mineral will be transported 400 km to a smelter, which will produce vanadium (V_2O_5) slag (2.4 ktpa), titanium (TiO_2) slag (106 ktpa) and pig iron (71 ktpa).

Metals trading company xxx will purchase the three mineral products ex smelter gate, for further railage to the export harbour and shipping to their end customers.

9.2. Implementation Strategy

The VTI Project will be implemented in three phases:

Phase 1 entails completing the mine design. For this purpose, a small, 2 tph pilot plant will be commissioned, where the mining, processing and smelting processes are tested over a 3-month period to confirm the project design parameters, generate heavy mineral feedstock for a smelt campaign and finalise the long-term customer offtake.

Phase 2 entails commencing with the construction, commissioning and operations of the mining, processing and toll smelt operations to produce vanadium slag, titanium slag and pig iron.

Phase 3 entails an expansion of operations. This could entail the construction of their own in-house smelter.

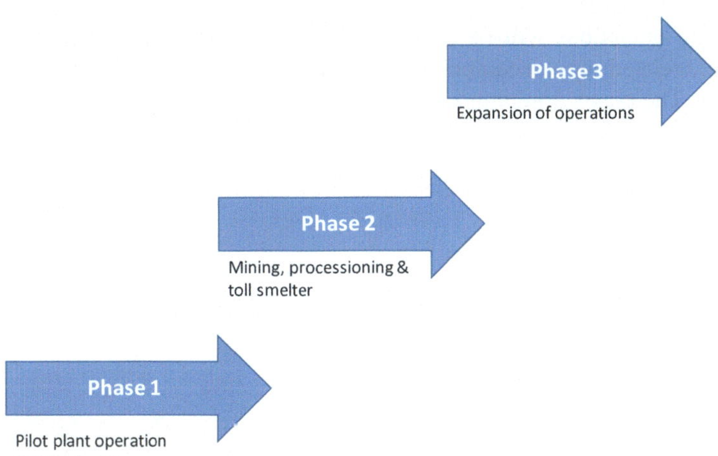

Source: The author

10. Mining

10.1. Mining Method

The mining operation entails a conventional hard rock opencast mine. Vegetation is removed and topsoil placed on a stockpile to be used later for rehabilitation. The exposed ore will be drilled by a hydraulic drill at a spacing of 3 m × 3 m and blasted. The blasted ore will be loaded and transported to the mineral concentrator. The waste rock will be stockpiled. The average strip ratio is 4:1. The waste product from the mineral concentrator will be loaded onto the trucks and transported back to the pit. Mining is to a depth of 120 m. The mining operations follow along the strike of the seam.

Source: The author

10.2. Mining Equipment

The mining operations will be outsourced to a reputable mining contractor. All mining and related support equipment will be provided and capitalised by the mining contractor. The planned mining equipment will entail:

3 × 85 ton excavators
12 × 40 ton ADTs
4 × drill rigs
2 × D10 dozers
4 × Front End Loaders 5m3
1 × Grader
1 × diesel bowzer
2 × water trucks
1 × Service truck
1 × Crane truck
6 × LDVs

10.3. Production Plan

The mining operation will produce 0.6 Mtpa of ore and 2.4 Mtpa of waste.

A mine plan has been developed for the first 10 years of the Life of Mine.

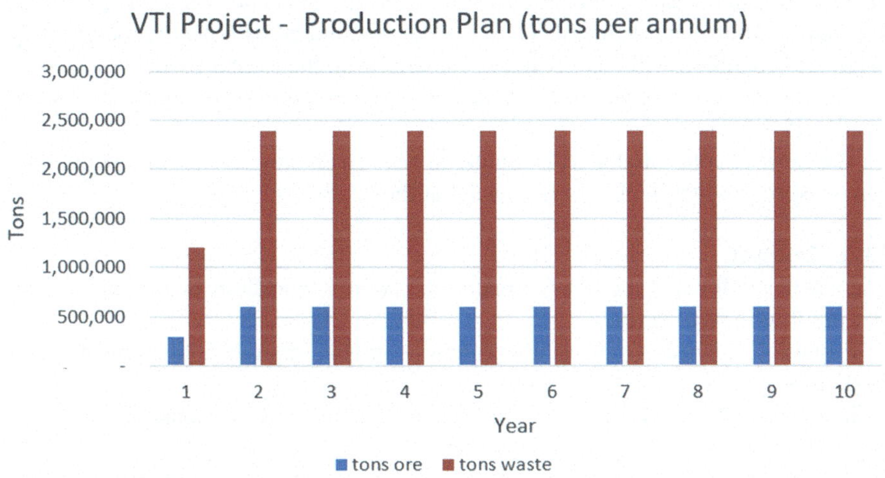

Source: The author

Rehabilitation will be carried out continuously over the life of mine.

10.4. Life of Mine

The mineable reserves are 20 Mt. At a production rate of 50,000 tpm, the Life of Mine of VTI Project is 33 years.

11. Mineral Processing

11.1. Mineral Concentrator

The Mineral Concentrator will produce heavy mineral concentrate via a series of mineral processing circuits for the smelter:

Crushing and Milling Circuit

The crushing and milling plant will be used to size the ore from the hard rock seam. It entails a series of crushing and milling circuits, including a grizzly and impact crusher, a jaw crusher, a cone crusher and a ball mill.

Primary Wet Circuit

The primary wet circuit entails a series of spirals to produce heavy mineral concentrate from the crushed and milled hard rock.

Mineral Separation Circuit

The heavy mineral concentrate will be upgraded in the mineral separation circuit to produce heavy mineral concentrate suitable for smelter feedstock. The mineral separation circuit uses a series of magnetic separation processes, including light intensity magnetic separation and wet high intensity magnetic separation processes to produce crude heavy mineral. The crude heavy mineral will then be roasted. Dry magnetic separation will recover heavy mineral from the roaster product.

The mineral concentrator will produce a 2,40,000 tpa of heavy mineral concentrate of 45% TiO_2.

The mineral concentrator will be built on site. The heavy mineral product will be conveyed to product bunker to be erected at the nearby rail siding.

11.2. Smelter

The smelter will smelt the heavy mineral to produce vanadium slag, titanium slag and pig iron in two stages.

In stage 1, the heavy mineral concentrate is molten to produce titanium slag and crude iron.

In stage 2, the crude iron is processed further to produce vanadium slag and pig iron.

VTI Resources has concluded a Toll Smelt Agreement with an existing smelter 400 km from the VTI mine site. This eliminates the capital cost to develop a new smelter.

The toll smelt agreement will last for a period of 10 years and is renewable for 5-year periods thereafter.

11.3. Pilot Plant

A pilot plant will be established to confirm the mine design parameters and the metallurgical processes. The pilot plant is a small-scale mineral concentrator which will produce heavy mineral concentrate for a smelt campaign, at 2 tons per hour.

Heavy mineral handling

Mineral concentrator

Toll smelter

Source: The author

12. Mine Infrastructure

The mine surface infrastructure at the VTI Project entails the following:

Access road	5 km access road from the main public road
Mine offices	Portable mine offices—provided by mining contractor and processing contractor
Mine access control and fencing	7 km of fencing
Weighbridges	2 × to measure haulers with heavy mineral concentrate leaving the mine site
A canteen and change house	To accommodate 300 personnel
A workshop	For the opencast mining fleet and mineral concentrator
Stores	For the mining contractor and processing contractor
Water supply	0.6 million m^3 per annum
Power supply	8 MW required for the mineral concentrator. Use of on-grid power and back-up generators. Solar solution to be investigated.
Mineral concentrator	100 tph concentrator with crushing, wet plant and magnetic circuits to produce heavy mineral concentrate
Tailings dam	500 m × 500 m × 40 m
Conveyor belts	2000 m
Product bunker	To handle 20,000 tpm heavy mineral concentrate for loading onto rail
Railway siding	To handle 20,000 tpm heavy mineral concentrate
Telecommunications	For mine and processing plants

Source: The author

The mine infrastructure forms part of the scope of work for the mining, processing, infrastructure and logistics contractors.

13. Market

13.1. Product and Prices

The VTI Project will produce vanadium slag, titanium slag and pig iron single metal trading company with whom an offtake agreement has been concluded.

Product	Uses	Buyer	Offtake price (FOT)
Vanadium slag	A new application of vanadium is vanadium redox batteries. Vanadium batteries are being developed for grid energy storage used for electricity generated by solar and wind power plants. This development has the potential to double the current use of vanadium.	Metals trading company xxx	$16,800/ton

(continued)

	Vanadium is used as an input in the steel industry as an alloy metal.		
Titanium slag 85% TiO_2	Titanium slag is used to produce TiO_2 pigment (for paint, coatings, plastics, paper, printing inks, rubber, textiles), titanium metals (for industrial applications, commercial aerospace, military, medical) and other uses for metallurgical fluxes and welding.	Metals trading company xxx	$650/ton
Pig iron	Pig iron is used as a leading input in the steel making process. The steel industry uses the material as a base product, which is then mixed with other metals. The demand for pig iron is related to the global demand for steel.	Metals trading company xxx	$350/ton

Source: The author

13.2. Customers

VTI Resources has secured an offtake agreement from reputable metals trading company xxx, who will purchase all the vanadium slag, titanium slag, and pig iron for the first 5 years of production at the agreed offtake price. This will significantly de-risk the VTI Project as the market is secured for 5 years. After 5 years, VTI Resources has the option to renew the contract or may elect to approach the end consumers directly.

13.3. Competitors

VTI Resources has the following primary competitor in the region:

Company xxx. A listed resources company which is developing a ferro-vanadium project which will produce of 1200 tpa of vanadium slag, 50,000 tpa of TiO_2 slag and 80,000 tpa of pig iron, located in xxx.

14. Logistics

14.1. Distance to Market

The VTI Project is located 400 km from the export port, from where the vanadium slag, titanium slag and pig iron will be shipped to the customers.

14.2. Mode of Transport

The heavy mineral concentrate will be transported from the mineral concentrator to the rail siding 10 km via road haulers. From there, the heavy mineral concentrate is transported 390 km via rail, to the toll smelter. The vanadium slag, titanium slag, and pig iron product produced from the smelter will be transported via rail from the smelter to the export harbour (30 km) on account of the metals trading company, for onward shipping to their customers.

14.3. Logistics Management

VTI Resources has appointed logistics contracting company xxx to manage the logistics scope of work of the VTI Project.

15. Human Resources

15.1. Organogram

The VTI Project will employ 236 employees. The majority (216) are the employees of the mining, processing and logistics contractors and will be sourced from the local region. VTI Resources will have 20 direct employees.

A project development team, ranging from 5 to 50 consultants and contractors, will be engaged during the mine design and construction phase by VTI Resources.

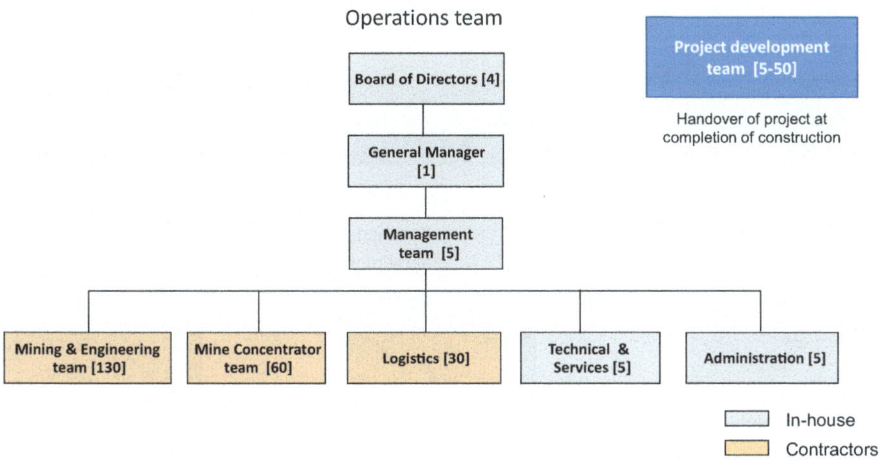

Source: The author

15.2. Mine Development Team

The mine development team will be responsible for the development and implementation of the VTI Project. The team includes experts in geology, mining, processing, smelting, engineering, operations management, mineral sales, logistics and contracting. The team has successfully developed titanium mines. The team members are:

- Exploration: xxx
- Feasibility: xxx
- Mining: xxx
- Processing: xxx
- Smelting: xxx
- Mine infrastructure: xxx
- Market: xxx
- Logistics: xxx

Once the VTI Project is developed, the development team will hand over the project to the operations team.

16. Contracting

16.1. Contracting Structure

VTI Resources will have the following key contracts with the following entities:

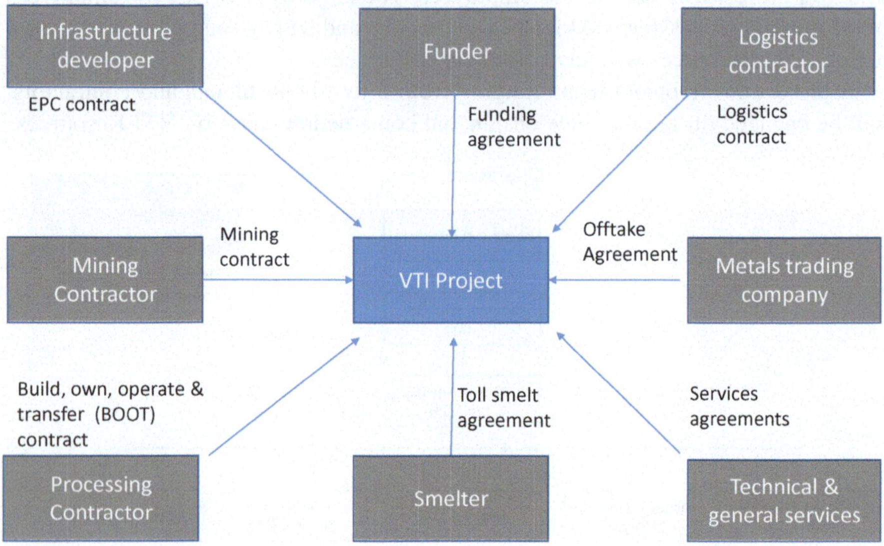

Source: The author

16.2. Key Suppliers and Contracts

Mining contractor	Company xxx A contract to establish operations and mine the ore. This includes providing mining equipment and manpower to service the contract.
Processing contractor	Company xxx A contract to build and operate the mineral concentrator. At the end of the contract period the mineral concentrator plant is transferred to VTI Project.
Smelter	Company xxx A toll smelter agreement to smelt the heavy mineral concentrate to produce titanium slag, vanadium slag and pig iron.
Infrastructure developer	Company xxx A contract to construct the required mine infrastructure: mine power, pollution control dams, mine tailings dam, access roads, mine access road, siding upgrade and heavy mineral storage facility.
Logistics contractor	Company xxx Contract to transport the heavy mineral concentrate to the smelter.

(continued)

Funder	Company xxx Funding agreements relating to the capital investment with financial terms and conditions.
Metals trading company	Company xxx Long-term mineral sales agreements with VTI Resources for titanium slag, vanadium slag and pig iron.
Technical and general services	Company xxx Contracts with service providers to provide technical and support services for the mining and processing operations.

Source: The author

17. Environmental Management Plan

17.1. Environmental Setting

The VTI project is situated in a bushveld setting. There are rare indigenous trees on the property. There is an informal settlement on the project area with several small subsistence farmers.

17.2. Environmental Management Plan (EMP)

The EMP has been completed by environmental consultancy xxx, who have engaged various specialists to conduct the fauna, flora, archaeological and geohydrological studies. Three consultative forums were held with affected parties.

The EMP was approved on xxx by the Department of Environmental Affairs. The environmental management plan entails:

- Removal and storage of topsoil in the mining pit, prior to mining
- No mining where rare indigenous trees and large trees have been identified
- Cultivation of seedlings for replanting
- Contouring, top soiling and re-grassing the mined mine out areas and waste stockpiles
- Water conservation and recirculation of industrial water
- Storm water management
- Waste management procedures

17.3. Rehabilitation Cost

The cost of rehabilitation has been calculated at $10 million. VTI Resources has lodged a rehabilitation bond with insurance company xxx. An allowance of $0.7/ton has been made for the rehabilitation of the mine and processing site.

18. Social and Labour Plan

18.1. Community

The community in which the VTI Project is located belongs to the regional xxx indigenous tribe. Several large-scale industrial farms are in close proximity to the mine site. There are approximately 30,000 people within a radius of 20 km from the project. The level of unemployment is high at 60%.

18.2. Local Economic Development

The local economic development initiatives selected by VTI Resources and approved in a series of consultations with community representatives are centred around development of technical skills to create employment opportunities.

- A commitment by VTI Resources to skill and employ a minimum of 70% of the employees from the community. This obligation is imposed on the mining and processing contractors.
- Identification of candidates from the community to skill in mining, engineering and mineral processing. A bursary scheme will be launched to enable five identified community members to go to university or technikons per annum.
- The installation of school IT facilities to enable the community to access online courses.
- The funding of off-grid solar installation for the community centre.

19. Financial Analysis

19.1. Input Parameters

The economic model was developed around the case of mining, processing and toll smelting of titaniferous ore to produce titanium slag, vanadium slag and pig iron for a metals trading company. The input parameters are as follows:

ROM (tons)	2,00,00,000
Mining method	Opencast
Production ore (tpm)	50,000
Stripping ratio	4:1
Production waste (tpm)	2,00,000
LOM (years)	33
Ore % V_2O_5	1%
Ore % TiO_2	20%
Ore % Fe	35%
Concentrator recovery	40%
Smelter recovery	74%
Price Vanadium slag $/ton	16,000
Price TiO_2 (85%) slag $/ton	550
Price Pig Iron $/ton	300
Royalties- revenue based	2%
Tax rate	26%

Source: The author

The economic model was developed for the first 10 years of the Life of Mine.

19.2. CAPEX Requirements

The CAPEX to develop the VTI Project is as follows:

Item	$
Mining boxcut	57,14,286
Mineral concentrator	1,50,00,000
Mine tailings dam	10,71,429
Pollution control dam	17,85,714
Bulk water supply	7,14,286
Logistics infrastructure	25,00,000
Fence	1,78,571
Mining operations establishment	9,28,571
Mine access road	7,85,714
Power supply	15,71,429
Project management	7,85,714
Contingency 10%	30,71,429
Total	3,41,07,143

Source: The author

19.3. Operating Costs

The operating costs have been calculated as follows:

Item	$/ton
Mining cost	2.1
Processing cost	10.0
Smelting	321.4
Transport	28.6
Management fee	2.1
Rehabilitation	0.7
	364.3

Source: The author

19.4. Project Cash Flow

The VTI Project cash flow is depicted below:

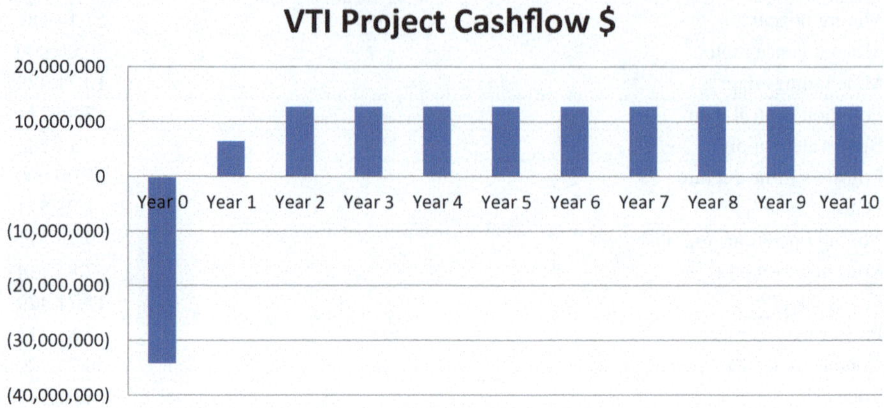

Source: The author

19.5. Financial Indicators

The VTI Project will generate a cumulative cash flow after tax of US$87.7 million over 10 years. The financial indicators Net Present Value (NPV) at a discount rate of 10% per annum, Internal Rate of Return (IRR) and Payback Period are:

Project analysis	
NPV (10%) ($)	3,53,27,442
IRR	31%
Payback period	4 years

Source: The author

19.6. Funding Requirements

VTI Resources is seeking funding of $34.11 million.

The gearing is 60%, resulting in the following debt and equity funding targets.

	%	$ mill
Total		3,41,07,143
Equity	40%	1,36,42,857
Debt	60%	2,04,64,286

Source: The author

19.7. Expenditure to Date by VTI Resources

VTI Resources has spent to date $2.5 million on the development of the project, which is broken down as follows.

Item	$
Acquisition	5,00,000
Exploration	5,71,429
Prefeasibility study	6,42,857
Permitting	35,714
EIA	2,85,714
Feasibility study	2,85,714
Management fees	2,14,286
Total	25,35,714

Source: The author

20. Risk Management Plan

VTI Resources is identified and ranked according to the technical, human and economic risks of the VTI Project and compiled the following to risk management plan for investors:

20.1. Technical Risks

Risk	Risk ranking	Risk management plan
Geological uncertainty	Low	The VTI titaniferous orebody is well explored
Mining	Low	VTI Resources has selected a reputable mining contractor to undertake opencast mining Extensive mine planning has been conducted by experienced mining consultants
Processing technology	Medium	VTI Resources is launching the project in with a pilot to confirm the processing methodology Appointment of reputable processing contractor Use proven technology
Infrastructure risk-scarcity of water	High	VTI has secured water supply upfront and intends to use grey water sources. Use of low water consuming processing methodology
Environmental risk	Low	Implement Environmental Management Report Ongoing monitoring
Contractors non-performance	Medium	Appoint reputable mining and processing contractors. Implement performance guarantees Secure step in rights
Lack of mine power	Medium	Investigate off-grid power solar diesel solution in long term

Source: The author

20.2. Human Risks

Risk	Risk ranking	Risk management plan
Workforce—industrial action	Medium	Ongoing communication and consultation with workforce in workplace forum meetings Ensure a minimum of 60% local employees, minimise external parties
Country politics	Medium	Political risk insurance

Source: The author

20.3. Economic Risks

Risk	Risk ranking	Risk management plan
Market forecast	Low	VTI Resources has concluded a long-term price hedging arrangement with a metals trading company
Financing	Medium	VTI Resources is seeking partnerships with reputable financiers who will conduct a due diligence on lenders
Completion risk	Medium	VTI Resources to secure completion guarantees from mining and processing contractors
Contractual risks	Low	VTI Resources has appointed a reputable legal firm to conclude the relevant legal agreements required to bring the VTI project into production

Source: The author

21. Project Plan

The activities and timelines to implement the VTI Project are outlined below. The duration from Prospecting Permit to operations is 3.5 years. Activities completed and those outstanding are outlined below in the Gantt bar chart below.

The feasibility study has been completed and the mining licence has been awarded.

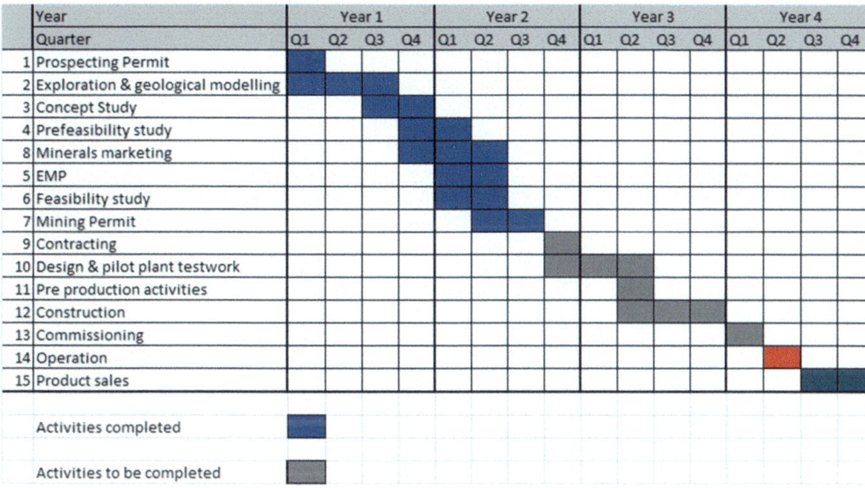

Source: The author

22. Conclusion

The development of the VTI Project is based on the following key considerations that make the project an attractive investment proposition:

- A well-explored, high-grade orebody
- An offtake agreement concluded with a metal's trader for the commodities vanadium slag, titanium slag, and pig iron where there is a strong demand
- Approved licences that enable mining to commence
- A strong mine development and operations management team

All these requirements can be met given a chance to implement this business plan. The team tasked to implement this plan has successfully developed and operated several titanium mining projects.

The VTI Project is a highly attractive project. It will generate a cumulative cash flow after tax of US$87 million over 10 years. The Net Present Value (NPV) of the project is $35 million and the Internal Rate of Return is 31%. The capital payback period is 4 years.

Raising Mining Capital: Best Practices

4

Contents

M. Seeger, *Mining Capital*, https://doi.org/10.1007/978-3-030-31225-1_4

4.1 Introduction to Raising Mining Capital

Having dealt with the characteristics of mining capital, the development of a mining business case and the marketing of the mining business case, we now move into the process of raising mining capital. Mining capital involves large sums of money, ranging from a few million dollars to tens and hundreds of million dollars, pending the business case. This process must be approached in a systematic and structured way, as investors have capital and are mandated to invest, but are highly risk averse

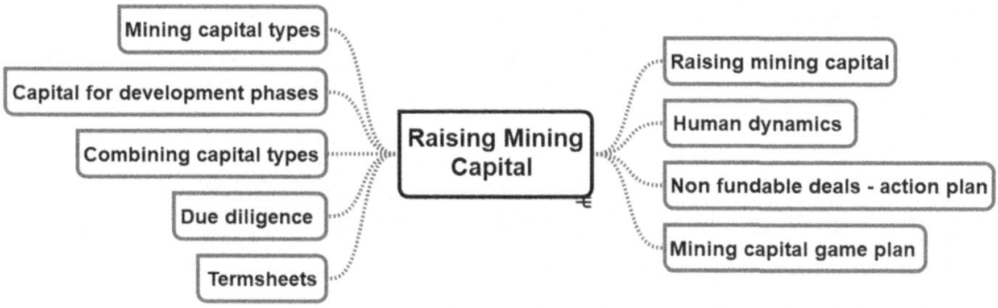

Fig. 4.1 Raising mining capital (Source: The author)

and more often than not decline to invest. But investments are made, and this happens when a systematic capital raising process is planned and executed successfully. This chapter covers the following topics as outlined in Fig. 4.1.

- Mining capital types: the different types of mining capital and the transaction structure behind each type.
- The different types of mining capital applicable to the project development stage, whether it is the exploration, feasibility, construction or operations phase.
- Combining different types of mining capital to finance projects.
- Due diligence: a detailed process into the technical, financial and legal aspects of the mining project. Only a successful due diligence process will result in investment. This is one of the most critical processes and must be well prepared.
- Mining capital term sheets: Generic term sheets and the transaction structure applicable to each mining capital type are provided. These can be used by the mine developer for their own project.
- Process of raising mining capital: A step-by-step process, from enquiry for mining capital to financial close, is discussed.
- Human dynamics: The human and psychology that influence decisions to invest capital into mining projects and operations.
- Non-fundable deal proposals: Why certain mining business cases cannot be financed and action plans to rectify.
- A mining capital game plan: A practical model for developers that can be applied to any mining project, at any development stage and within any commodity, seeking capital.

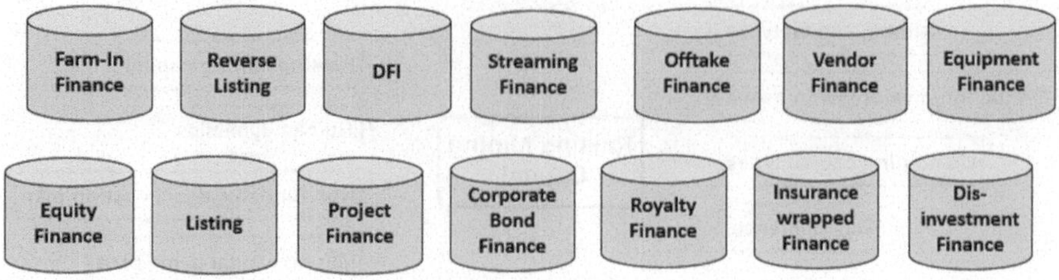

Fig. 4.2 Mining capital types (Source: The author)

4.2 Mining Capital Types

Different types of capital are required for the different project stages, and the mining project developer needs to understand this throughout the project development stages. The following types of capital are available for mining projects [1, 5, 8, 9, 10, 11, 53] (Fig. 4.2):

The types of mining capital are outlined in terms of:

- A definition of the type of capital
- The transaction structure, depicting at what level the investor enters the project—whether at Holdco (company shareholder) level or at SPV (project company) level
- Notes pertaining to the mining capital type
- An example

4.2.1 Equity Finance

Definition

Equity finance is funding that is provided by an investor interested to invest into a project and acquire shares in it. The investor participates in the increase in the value of the mining project. The investment is used for the ongoing financing of the mine development programme, which is either in exploration, development or expansion of the project. The investor invests capital into the SPV in exchange for shares in the project, with Holdco providing warrantees (Fig. 4.3).

Transaction Structure

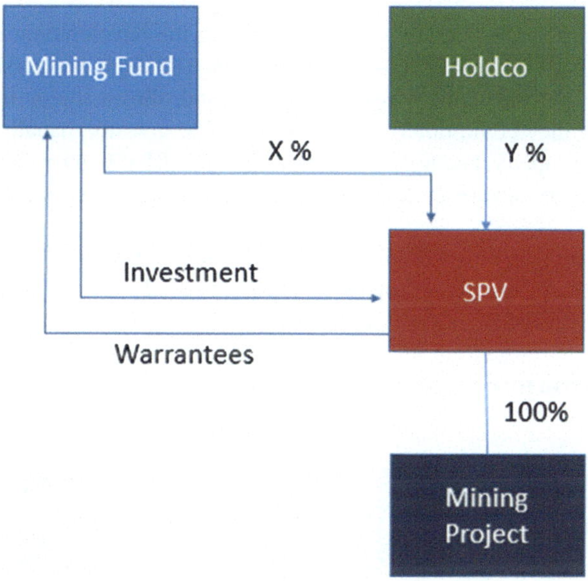

Fig. 4.3 Transaction structure—equity finance (Source: The author)

Notes

Equity capital can be provided by private investors, a mining fund, larger mining companies or commodity traders.

Example

- *Copper project*
- *Requires $15 million to assist with mine development*
- *Investor: mining fund*
- *Investment: $15 million*
- *Shares: 150 million new fully paid ordinary shares*
- *Share price: $0.10/share*

4.2.2 Farm-In Finance

Definition

Farm-in funding entails a mining investor funding the development of the mining project in a staged manner. This form of funding is a strategic partnership between a junior mining company and a larger company and allows junior miners to gain funding, which they could otherwise not access. The investor, usually a mid-tier or major mining company, acquires equity in the project SPV by providing investment tranches into the project (Fig. 4.4). The typical investment is stages are as follows:

- Investment tranche 1: to complete the exploration campaign
- Investment tranche 2: to complete the feasibility study
- Investment tranche 3: to develop the mine

Transaction Structure

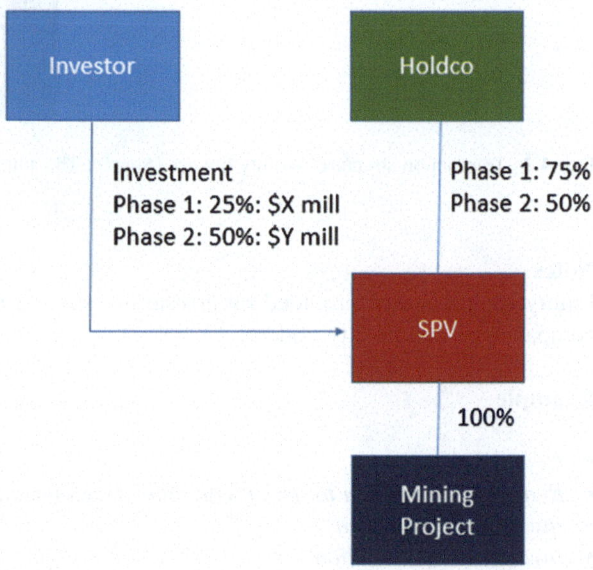

Fig. 4.4 Transaction structure—farm-in finance (Source: The author)

Notes

The investor commits a minimum fixed expenditure over an agreed period.

The investor has the right to spend further at its own election or the obligation to incur further outlay if certain targets are achieved.

The share price for the different stages may be predetermined for each investment stage.

Example

- *Investor: Major mining company*
- *Holdco: Listed junior mining company*
- *Project: Platinum and nickel exploration properties*
- *Deal: Major mining company agrees to provide a maximum of $3 million that will be used to fund a 3-year work programme. Upon completion, the major mining company will earn a 50% interest in the project.*

4.2.3 Listing

Definition

Listing entails the developer taking a company public onto a stock exchange in order to raise equity capital. A listing on a stock exchange lends a mining company credibility and profile, makes investors more likely to purchase its shares and enhances the ability to raise capital. Companies wishing to issue their shares on a stock exchange must meet specific listing requirements and continue to do so for as long as they are on the exchange.

Transaction Structure (Fig. 4.5)

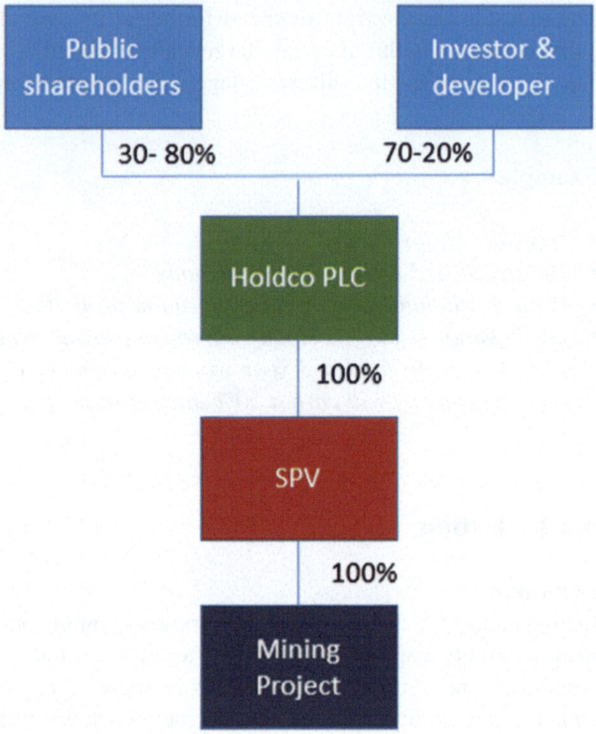

Fig. 4.5 Transaction structure—listing (Source: The author)

Notes

Listing process and key considerations

The listing of a mining company varies from stock exchange to stock exchange, but in general, it is as follows: mining company intending to list appoints corporate financial advisor (also called nomad or sponsoring broker) who will coordinate the team (legal advisor, mining consultants, underwriters, and retail broker) and liaise with the stock exchange and securities commission to complete the listing process (Fig. 4.6).

Identifying investors

The corporate finance advisor compiles an information memorandum and identifies institutional investors either directly or via brokers and underwriters. Roadshows are organized requiring the developer to give the investor presentation to multiple types of investors.

Disclosure of documentation

Independent mining consultants undertake an independent valuation of the mining assets. The auditor compiles management accounts and annual financial

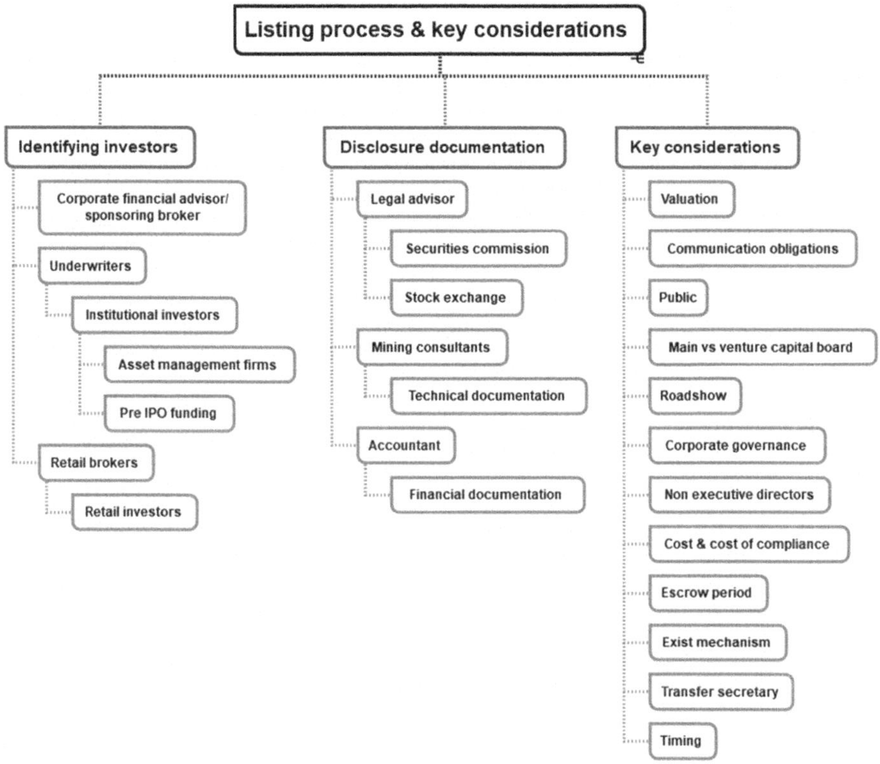

Fig. 4.6 Listing process and key consideration (Source: The author)

statements. The lawyer undertakes a due diligence on company secretarial documentation, agreements, and licences.

Institutional investors

An institutional investor is an organization that trades securities in large enough share quantities or dollar amounts that it qualifies for preferential treatment and lower commissions. Examples of institutional investors include pension funds and life insurance companies. A mining company will want to focus on institutional investors as they invest larger amounts of capital in mining stocks and understand the industry.

Asset management firms

Listed companies are attractive investment destinations for fir asset management funds, which are managing pension funds and can only invest in public companies.

Pre-IPO funding

The information memorandum is circulated to institutional investors prior to the listing for a pre-IPO funding. Should the investment story be attractive, equity capital is raised prior to the listing with shares issued at discounted prices. Once the listing is made public, retail investors may acquire shares in the Holdco PLC.

Table 4.1 Listing—role players and responsibilities

Advisor	Responsibility
Corporate financial advisors/ sponsoring broker	The role of the corporate financial advisors is that of marketing the company and its shares to potential investors and the investment community, while being a liaison between the company and the stock exchange and ensuring compliance by the company of all stock exchange rules
Independent mining consultant	The independent mining consultant undertakes an independent technical valuation of the mining asset. This is usually a Competent Persons Report in alignment with one of the accepted codes: SAMEREC, JORC, NI 43–101
Legal advisor	The legal advisor will review the information memorandum and company secretarial information and ensure that the mining company follows the stock exchange rules
Accountant	The accountants ensure that the mining company has audited annual financial statements, and monthly and quarterly management accounts specific to the listing
Transfer secretary	The transfer secretary maintains the company's register of shareholders, settles dividends and distributes company documentation to shareholders

Source: Authors' table based on experience of listing a junior mining company

Retail investors

Retail investors are individual investor who buy and sell securities for their personal account and invest small amounts.

Advisors

A company seeking to list on a stock exchange will require a series of prescribed advisors, typically a sponsoring broker, lead financial advisor, legal advisor, accountants, PR agent, and transfer secretary. The advisors and their responsibilities are outlined in Table 4.1.

Technical documentation

This is the Competent Persons Report on the mining asset(s) of the Holdco to be listed.

Financial documentation

Financial documentation required for a listing is the auditor's report, which includes annual financial reports, management accounts, budgets and cash flow forecasts.

Valuation

The valuation of the mining company is required to determine the value of the stock. It must be undertaken by independent mining consultancy.

Communication obligations

Once a mining company (Holdco PLC) is listed, it must communicate to the market on a quarterly basis the progress on the development of the projects. On an annual basis, the annual financials are compiled by a reputable auditing firm.

Public

The public is defined as any entity which owns less than 10% of the issued shares and is not related to the company and its directors.

Main board vs venture capital board

Many stock exchanges have two types of boards: a main board and a venture capital board. The listing rules for a venture capital board listing are less stringent than those of a main board (e.g. no audited profit history, less shareholders, less capital raising requirements). The listing requirements for a main board usually require more shareholders, a satisfactory profit history for multiple years, a minimum of shares in issue and a minimum initial issue price of securities.

Roadshow

The roadshow is a significant element in the listing process. This is where the mining business case is marketed. It entails a series of investor presentations to institutional investors, and thereby motivating them to invest in the listed stock. The roadshow is organized by the sponsoring broker.

Corporate governance

Corporate governance is essential for listed companies. It is defined as the system of rules, practices and processes by which the mining company is directed and controlled. It encompasses anything pertaining to management, from action plans and internal controls to performance measurement and corporate disclosure. The board of a company intending to list needs to undergo training in corporate governance.

Non-executive directors

A listed mining company is required to appoint non-executive directors to the board to ensure compliance and corporate governance and provide an independent but non-executive viewpoint in the decision-making process.

Costs and cost of compliance

The listing costs vary from stock exchange to stock exchange. The cost of compliance for quarterly and annual reports, independent audits, policies and host country specific procedures can be significant specifically for junior mining companies with no cash flow generating mining assets.

Escrow period

The mine developer may not sell his/her shares for a certain period, which is termed the escrow period, usually 2 years. This is to ensure that the mine developer remains committed to the project.

Exit mechanism

Listings present an exit mechanism for the mine developer and investors in the company. Apart from the developer/sponsor, an investor in shares may sell or buy shares at any point in time.

Timing

The timing of a listing is dependent on the magnitude and complexity of the mining deal (Fig. 4.7). It can range from 8 weeks to up to a year. A listing milestone chart for a company raising $20 million is outlined in Fig. 4.2.

Many failed listings in mining

Many junior mining companies listed on stock exchanges during the commodity boom phase. It was a period of easy access to mining capital. Many mining business cases and investment propositions were highly speculative and flawed, and developers failed to deliver on expectations, leaving investors disappointed. In many instances, the investors in such listed mining stocks lost significant amount

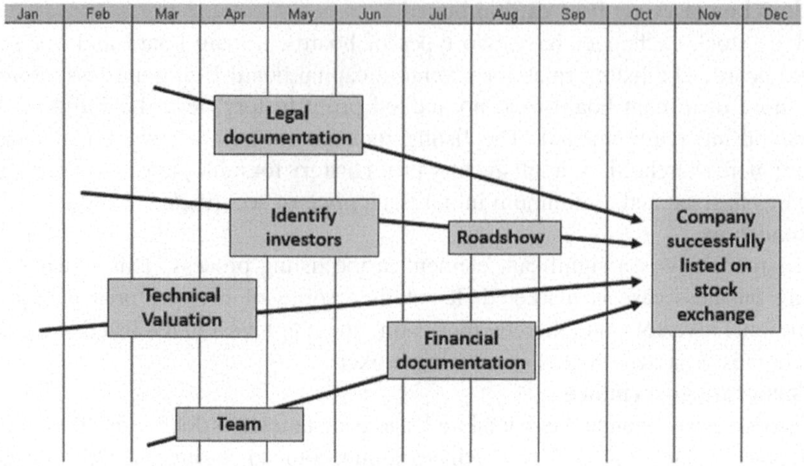

Fig. 4.7 Example listing process and timelines (Source: Seeger, M. Development of a Strategic and Tactical Game Plan for Junior Mining Companies [44])

of investment. This, in turn, has put off many funds into investing in listed junior mining stocks.

Not all mining companies and projects are suitable for listing

Some mining companies and projects and companies are not suitable for a listing. Listing relies on public perceptions for purposes of valuating a company and share price appreciation. An example is mining companies in the coal sector, a sector with a negative public perception of contributing to global warming. No matter how well a coal company will perform in terms of deliverables, the share price will not move significantly if public perception is against it, with no fault of the mine developer.

Equally, a company with a strong entrepreneurial flair should consider whether listing is the ideal way to raise capital. Managing a listed company requires strict corporate governance, frequent structured reporting to the market, a representative corporate image, amongst the many obligations. These obligations can stifle the entrepreneurial spirit of junior mine developers.

Example

- *Mining company requires $5 million to conduct the feasibility study and testwork for a phosphate project.*
- *Process:*
  - *Develop mining business case, investor presentation business plan*
  - *Complete Competent Persons Report*
  - *Establish data room with project data: technical, financial, legal*
  - *Identify and mandate sponsoring broker*
  - *Appointment of accountants and lawyers*
  - *Due diligence exercise*

  - *Payment of admissions fees*
  - *PR coverage*
  - *Roadshow*
- *Pre IPO funding*
- *Listing on stock exchange: issue 50 million shares at $0.10/share.*
- *Receipt of funds*
- *Payment of commissions*
- *Reporting*

4.2.4 Reverse Listing

Definition

A reverse listing entails shareholders of a mining project (Holdco) vending in its shares in the SPV, which owns the project, into a listed company PLC. In exchange, the Holdco shareholders are issued shares in the listed company PLC.

Transaction Structure (Fig. 4.8)

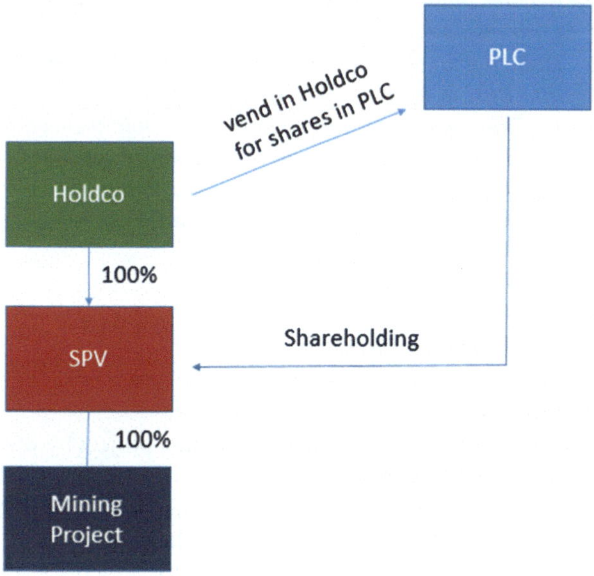

Fig. 4.8 Transaction structure—reverse listing (Source: The author)

Notes

- There are a series of listed junior mining companies on stock exchanges around the globe that are searching for projects to vend into their portfolio. Many juniors have listed based on perceived solid projects. Should they not materialize however, they will have to be replaced with viable projects to keep investors satisfied. This is where reverse listing plays a role.
- Reverse listing is a suitable method for financing of exploration projects.
- Prior to reverse listing, the project is valued by a Competent Person. The project value is compared to the market cap of the listed company and this provides a basis for negotiating the value of shares of the Holdco shareholders in the PLC.
- A reverse listing can be a cheap way to gain high profile and access to the capital raising capabilities of a company listed on a stock exchange.
- The downside: Holdco shareholders must give away control and enter the strict corporate governance environmental of a listed company.

Example

- *Mining company Holdco requires $2 million to conduct the feasibility study of a gold project.*
- *A listed junior mining company (PLC) is identified with $0.5 million in cash reserves and no project.*
- *Holdco shareholders vend their gold project SPV with the prospecting licence into PLC in exchange for shares in PLC.*
- *PLC announces the acquisition of the gold project. $0.5 million is used as a first payment to commence with the feasibility study.*
- *PLC raises a further $3 million via a rights issue.*
- *The proceeds are used to pay for the feasibility study and business development.*

4.2.5 Project Finance

Definition

Project finance is the funding of a mining project by way of a loan that is secured by the project. The loan is paid back from the cash flow generated by the project, and the bank has no recourse to the balance sheet of the shareholder. The project's assets, rights and interests are held as collateral. Holdco must provide completion guarantees to the bank. Project finance is suited for high CAPEX mining projects and entails long-term loan agreements. Project financing is available to mining projects that have secured all licences, completed the feasibility study, secured an offtake agreement for the commodity, have a strong mining team and received quotations from major EPC and O&M contractors.

Transaction Structure (Fig. 4.9)

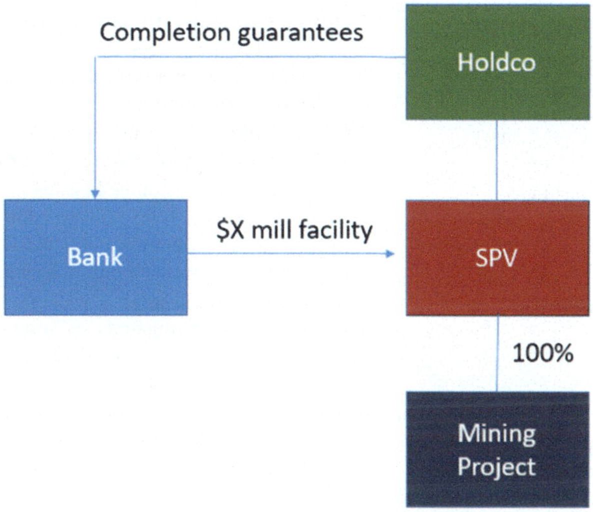

Fig. 4.9 Transaction structure—project finance (Source: The author)

Notes

Project finance is predominantly for mid-tier and major mining companies

Project finance requires an extensive technical, financial and environmental due diligence study. Project finance is applicable to projects where the feasibility study has been completed, the mine design has commenced, and the next step is the construction phase. This is the space of advanced mid-tier mining companies and major mining companies.

Requirements by the banks

Banks providing project finance consider the following in assessing the transaction (Table 4.2):

Financial covenants

The financial model of the project must be tested with respect to the following key financing criteria, which are integrated in the term sheet (Table 4.3):

Example

- *Mining company requires $20 million to develop a gold mine, after completion of the feasibility study.*
- *Lender: commercial bank*
- *Facilities: senior debt facility and standby debt facility*
- *Ranking: senior secured*
- *Interest: 9%*

Table 4.2 Project finance—bank consideration and description

Bank consideration	Description
Debt to equity ratio	The amount of equity the mine developer is contributing to the development of the project. Banks require the presence of a strong equity shareholder to lean on if further capital than planned is required and share the capital responsibility
Host country dynamics	The banks willingness to provide financing will depend on the jurisdiction in which the project is incorporated
Insurance	The bank will require comprehensive insurance for the mining project, such as third-party liability, construction risk and delay in start-up. In political high-risk countries, political risk insurance is required to cover expropriation of assets and political violence
Title risk	The risk relating to licences is at the forefront of the project financier's due diligence. Often, minerals are vested with the host country and the central government issues and the mining company will obtain a licence to mine the minerals in question at the discretion of the host government. Lenders will perform due diligence to ensure that such licence is valid and in what circumstances it can be revoked. In many countries, the host government will retain a free carried interest in the project or require royalty payments
Conditions precedent to drawdown	Conditions precedent to drawdown in a project financing loan agreement will include the completion of the feasibility study, obtaining all required licences from government, obtaining an offtake agreement and contracts for the development of mine infrastructure, mining and processing
Step in rights	The bank will want the right to step in place of the borrower in the event of default
Security	The bank will want to take security over the project assets
Ranking of the debt facility	The debt is ranked senior and must be repaid first if the company goes out of business
Independent technical consultant	An expert is appointed by the bank to undertake a high-level review of technical information of the project during the due diligence process
Completion test	The bank will require a completion test when the mine has been constructed. This test is defined as the passing of technical, operational and economic completion tests. These are developed in conjunction between the borrower, the lender and the independent technical consultant, who will be appointed by the lender to verify the satisfactory results of these tests. Once completion tests are concluded, loans become non-recourse to the sponsors and shareholders
Recourse	In frontier regions, the lender may seek recourse to the shareholders via form of guarantee or security agreement during the construction and commissioning phase, to reflect the increased risks associated with mine development. Once completion tests are concluded, loans become fully non-recourse to the sponsors and shareholders

Source: The author, based on working with project finance term sheets for mining projects

Table 4.3 Project finance—financial covenants

Financial covenants	Definition
Cash flow available for debt service (CFADS)	The available cash (revenue minus operating costs and tax) to service the debt on a monthly or quarterly basis
Debt service cover ratios (DSCR):	The ratio of cash available for debt servicing to interest, principal and lease payments in any given period
Loan life cover ratios (LLCR):	This is the net present value of cash flow available for debt service, which is measured up to the maturity of the debt tranche, and provides an estimate of the credit quality of the project from a lender's perspective
Reserve life (tail) cover ratios (RTCR)	The ratio of the saleable commodity reported in the ore reserves remaining to be mined at the scheduled maturity of the facility to the total commodity reported in the ore reserve statement

Source: Authors' table based on mining financing term sheets

- *Maturity: 7 years*
- *Payment holiday: 6-month repayment holiday after which a 12-month capital repayment holiday*
- *Security:*
  - *Pledge of SPV shares*
  - *Mortgage over the land housing the mineral licence*
  - *General notarial bond over all movables*
  - *Guarantee from Holdco*
  - *Cession of the collection account;*
  - *Cession of specific assets of Holdco*

4.2.6 Development Finance

Definition

Development finance is provided by Development Finance Institutions (DFIs). DFIs invest in mining projects at the feasibility study, construction or operational stage. Commercial banks may not be able to fund mining project which are in high-risk frontier countries. However, DFIs can provide credit in the form of loans, equity stakes and risk guarantee instruments to mining companies investing in such high risk developing or frontier countries. DFIs will require strict compliance with social and environmental standards. DFIs have a mandate to develop an industry and create jobs in their host country. Once a DFI has invested in a project, future investors are likely to have greater confidence in the project.

Transaction Structure (Fig. 4.10)

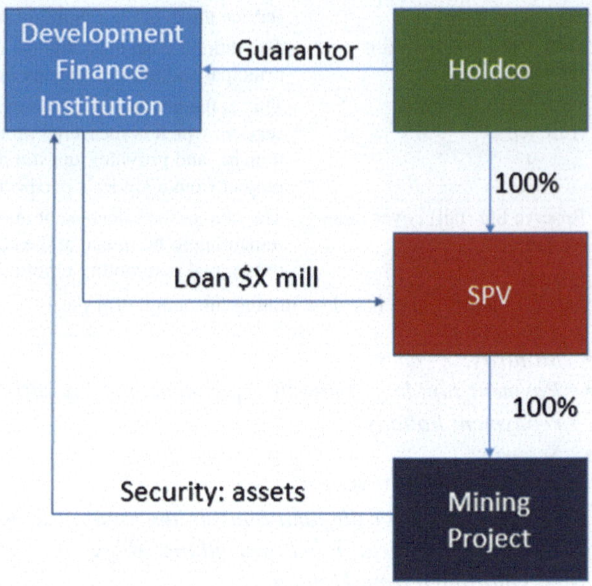

Fig. 4.10 Transaction structure—development finance (Source: The author)

Notes

Due diligence

The due diligence involved in development finance can be significant and take up to 3 months.

Funding structures

DFIs provide a variety of investment instruments such as senior debt, subordinated debt, equity and convertibles.

Environmental standards

DFIs are under public scrutiny over their mining investments and apply stringent environmental and social standards. Consequently, DFI's focus extensively on the due diligence process prior to entering a loan and include suitable protections such as representations, warranties, covenants and indemnities in their term sheet. Many DFIs require the mine developer to comply with Equator Principles outlined in Sect. 2.9.

Timing

It can take up to 12 months to secure development finance.

Example

- *Potassium project, South America*
- *Requires $70 million to develop mine*

- *Lender: Development finance institution*
- *Borrower: SPV*
- *Guarantor: Holdco*
- *Amount of facility: $70 million*
- *Interest rate: 9.25% pa*
- *Tenor: 60 months*

4.2.7 Corporate Bond Finance

Definition

Corporate bonds are a form of debt financing. A corporate bond is a debt security issued by a mining company and sold to investors.

Corporate bonds are issued in blocks, typically ranging from $100–100,000 in value, and have a standard coupon payment structure. The investor is the owner of the bond and receives interest from the issuer until the bond matures. At that point, the investor can reclaim the face value of the bond.

The backing for the bond is usually the payment ability of the mining project, which is earned once the mine is operational. The mining project's physical assets are used as collateral for bonds. A trustee manages the issuing of bonds and payments.

Transaction Structure (Fig. 4.11)

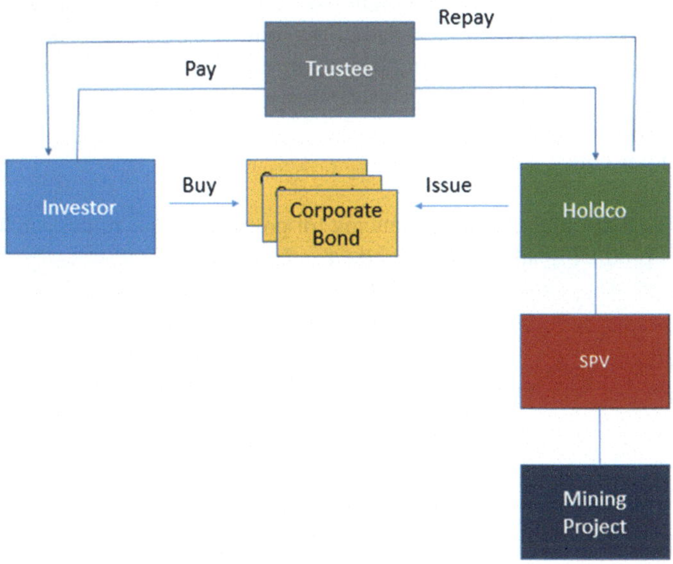

Fig. 4.11 Transaction structure—corporate bond finance (Source: The author)

Notes
With bonds, investors only earn interest rather than profits.

Example

- *Titanium project*
- *Funding requirement: $30 million to develop the first phase of the mine*
- *Financial instrument: secured redeemable bonds*
- *Holder: investor*
- *Period: 5 years*
- *Interest: 8% pa*
- *Value per bond: $100,000*
- *Number of bonds: 300*
- *Value of issued bonds: $30,000,000*

4.2.8 Streaming Finance

Definition
Streaming finance is a popular form of mine financing for mines producing precious metals. The investor provides an upfront payment to the mining company in return for the right to a percentage of production. This is usually for life-of-mine from a mining project where the underlying asset is precious metals. The investor will be looking for steady and stable production over a long period. Stream payments are ongoing on receipt of physical metal. Streaming arrangements do not lead to shareholder dilution. Each streaming deal is different, but typical features include:

- Upfront cash payment
- Ongoing per ounce payments

Streaming is a popular form of financing for precious metals mines, such as a gold mine that is producing silver as by-product.

Transaction Structure (Fig. 4.12)

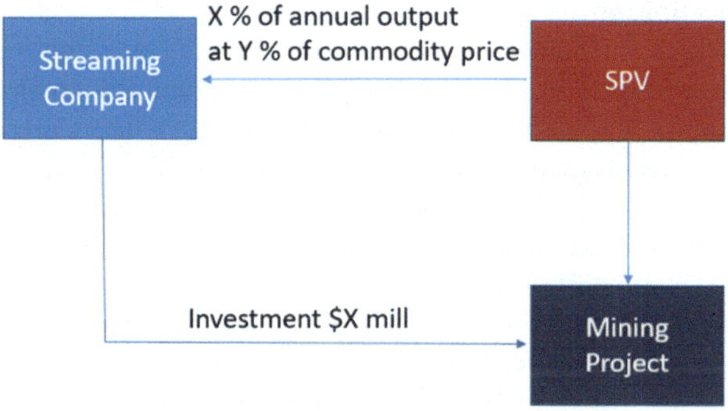

Fig. 4.12 Transaction structure—streaming finance (Source: The author)

Notes

- Streams involve deliveries of metal or metal credits.
- The main points of negotiating a streaming agreement will be:
  - the amount of the upfront payment
  - the purchase price
  - the term
  - the percentage of total production
  - the method of delivery.
- Stream holders stand with other creditors in cases of financial distress.
- The mining company may negotiate an option to buy back a portion of the stream and may want other protections in the event of an interruption of production.
- Metal streaming agreements are covenant-light compared to debt financing.
- The mining company retains the option of securing additional funding from other sources.
- The investor is dependent on production, is not a shareholder and has no control over management of the mining company. It will place significant focus on technical due diligence of the project.

Example

- *Project: Gold project*
- *Developer: Gold–silver mining company*
- *Funding requirement: $60 million*
- *Streaming deal:*
  - *Streamer pays gold mining company $60 million*
    Right to purchase 12% of the gold production at $450/oz
    Right to purchase 22% of the silver production for $5/oz
  - *Duration: until capital plus interest is repaid*

4.2.9 Royalty Finance

Definition

Royalty finance is similar to streaming finance and a popular form of alternative mine financing. The investor provides an upfront payment to the mining company in return for a share in the share of the project's future revenue for the mine life. Forms of royalties are:

- Gross Royalties: These royalties are based on total revenue from the sale of commodity
- Net Smelter Return (NSR): This royalty entitles the holder to an interest in the proceeds paid to the miner by a smelter.
- Net Profit Interest: This royalty is based on profit after deduction of costs of production.

Transaction Structure (Fig. 4.13)

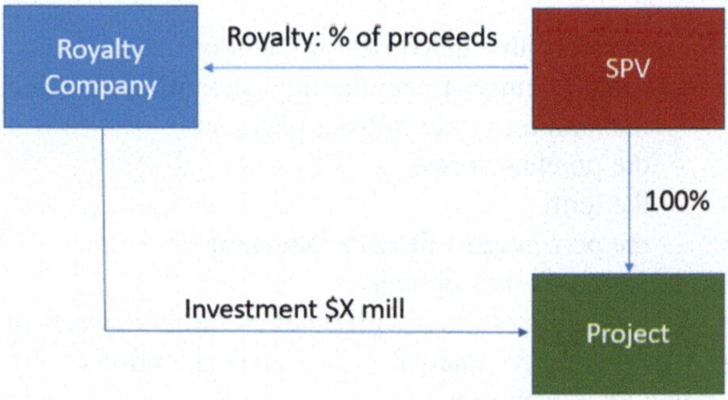

Fig. 4.13 Transaction structure—royalty finance (Source: The author)

Notes

- Royalties are secured on the mining property and bind any purchaser of the property in the event of a transfer.
- A royalty agreement needs to consider the local jurisdiction as not all jurisdictions allow the mining property to be ceded to the investors as a guarantee.
- Royalties are paid in cash.
- Royalties can be on all minerals.

Example

- *Project: Gold project*
- *Mine developer: Gold mining company*
- *Investor: Royalty company*
- *Funding requirement: $10 million*
- *Royalty deal:*
  - *Royalty company pays gold mining company $10 million*
  - *Gold mining company pays Royalty company 2.4% of the gross value of recoverable gold (Net Smelter Return).*
  - *Duration: entire term of the mineral lease*

4.2.10 Offtake Finance

Definition

Offtake finance or pre-production finance is a form of financing for mines producing bulk commodities, such as coal or iron ore. The offtaker/investor, who is a minerals commodity trader, prepays the mining company SPV to deliver mineral product at a delivery point, 1–3 months prior to delivery.

Offtake finance does not lead to shareholder dilution. Each offtake finance deal is different, but typical features include:

- Quantity and quality definitions for the mineral product
- Purchase price
- Prepayment terms
- Delivery point

Transaction Structure (Fig. 4.14)

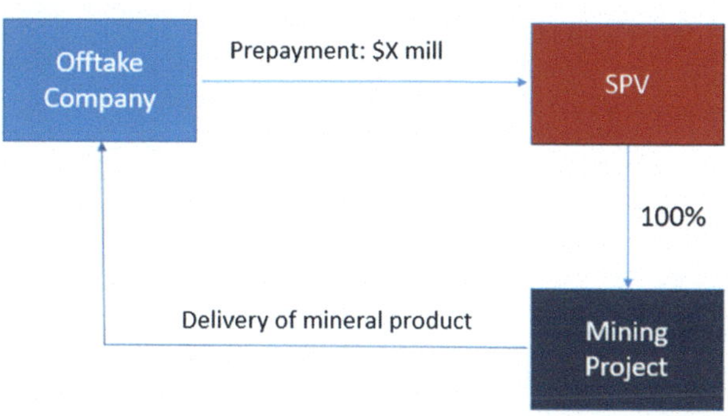

Fig. 4.14 Transaction structure—offtake finance (Source: The author)

Notes

Offtake finance is applicable for bulk commodity mining projects, such as iron ore, coal, bauxite or phosphates that are operational or about to start with operations. Penalties can be severe if the mining company fails to deliver the required quantity and quality of mineral product per contract specifications. Hence the mining operations need structured with skills, expertise, systems and equipment prior to entering into an offtake agreement.

Ongoing communication between the mining company and the offtaker is key to success.

The risk and title pass from the mining company to the offtaker as soon as the mineral product is loaded onto trains at the delivery point.

Example

- *SPV: Coal company*
- *Investor: global coal trader*
- *Product: thermal coal*
- *Quantity: 100,000 tons per month*
- *Quality: CV 26 MJ/kg*
- *Prepayment: $6 million paid 3 months prior to delivery of coal for 1 month's production of coal*

4.2.11 Insurance Wrapped Finance

Definition

Insurance wrapped finance entails and insurer approving the mining business case and insuring the funder of the project, i.e. the banks. The insured banks pay the supplier of the mining equipment for these to provide the goods to the mine. The SPV repays the equipment supplier, who in return repays the bank. In the event of a default, the insurer repays the bank.

Transaction Structure (Fig. 4.15)

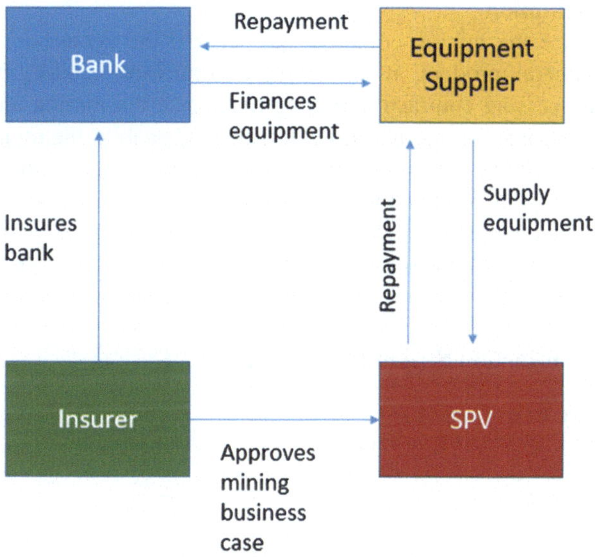

Fig. 4.15 Transaction structure—insurance wrapped finance (Source: The author)

Notes

- This is a new type of mining finance and particularly suited to providing mining to projects in politically instable host countries.
- If the bank is insured, it will fund the transaction.
- In the event of default, the insurer repays the bank.
- The foundation of this type of funding is a solid mining business case.

Example

- *Gold mine*
- *Funding requirement: $25 million to purchase a gold processing plant*
- *Borrower: SPV*
- *Insurance company: insures business case*
- *Bank: provides debt facility to purchase equipment, value $25 million*
- *Equipment supplier: Mineral processing company appointed to design and build the gold processing plant under a EPC contract. Value of contract: $25 million.*
- *SPV repays the equipment supplier who in turn repays the bank.*

4.2.12 Contractor Vendor Finance

Definition

Once the mining company has secured the licence to mine and a market for the mineral product, it can explore contractor vendor financing from mining and processing contractors willing to finance the mining and processing equipment in exchange for a long-term contract. This enables the mine developer to fund organic growth without significant capital outlays. Large contracting companies are willing to explore vendor financing for their clients to ensure that they have a pipeline of projects. The mine developer must enter a long-term contract with a mining and/or processing contractor under a Build + Own + Operate (BOO) Contract, where the vendor finance suppliers recover their capital via per ton operating costs.

Transaction Structure (Fig. 4.16)

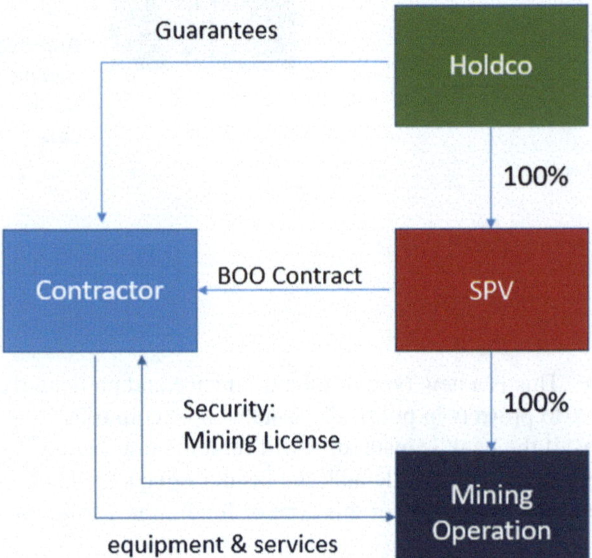

Fig. 4.16 Transaction structure—contractor vendor financing (Source: The author)

Notes

The mine developer must ensure that:

- The mining licence is intact
- The geological model is available
- There is a guaranteed market for the mineral product, with short-, medium- and long-term contracts
- Duration of contract: until equipment capital is repaid.

The mine developer usually needs to pay the contractor a fixed monthly fee and a variable fee $/ton.

Example

- *Dolomite mine*
- *Requires $7 million to expand operations*
- *Contractor funder: mining contracting company*
- *Borrower: SPV*
- *Contract type: Build, Own, Operate*
- *Scope: Turnkey mining solution, 5 years*
- *Value of contractor vendor financing for mining equipment: $7 million*
- *SPV to pay contractor fixed fee ($600,000/month) and variable fee ($15/ton)*

4.2.13 Equipment Finance

Definition

Mining and processing equipment can be financed by the contractor with an underlying contract (Fig. 4.17). The cash flow generated from the mining operation enables the payment of monthly instalments and interest to repay the equipment. The SPV needs to award the contractor a contract of significant duration to repay the equipment. With the contract in hand and adequate financials, the contractor can secure finance to purchase the required mining or processing equipment (Fig. 4.18). Often equipment suppliers, who are backed by commercial banks, require a 3-month instalment deposit and a surety. Typical finance structures include:

- Finance and operating leases
- Sale and leaseback
- Hire purchase
- Asset-based inventory financing

Transaction Structure

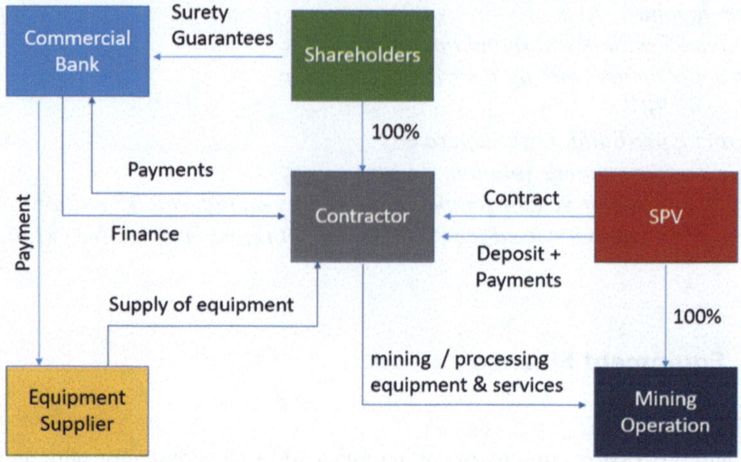

Fig. 4.17 Transaction structure—equipment financing (Source: The author)

Notes

This form of financing enables the mine developer to fund organic growth without significant capital outlays.

The borrower will be required to sign up for maintenance, repair and insurance covenants.

The security of this financial transaction can range from the assets in the project to the balance sheet of the shareholders of the borrowing company, the contractor.

Not all contractors are willing to provide security.

The borrower may be able to enter into a buy-back arrangement with the supplier, in which the supplier agrees to repurchase equipment at the end of the project.

The supplier may enter into a lease-to-buy arrangement with the mining company, where the supplier leases the equipment to the borrower and the project eventually buys the equipment at a discounted price. In this instance, the supplier takes security over the equipment.

Example

- *Mining contractor*
- *Seeking equipment finance for 2 × dump trucks*
- *Value: $599,000*
- *Interest: 15.5% pa*

- *Number of instalments: 48*
- *Amount per instalment: $12,604*
- *Securities: provided by shareholders of contractor*

4.2.14 Disinvestment Finance

Definition

Divestment finance entails the mining company selling non-core mining assets to raise capital to fund other projects and improve their balance sheet. Majors mining companies often dispose of smaller, mature mining assets that lack sufficient scale. Such asset disposals provide junior and mid-tier mining companies an opportunity to acquire mature but well-managed operations, ring-fenced in an SPV, and diversify their asset base.

Transaction Structure

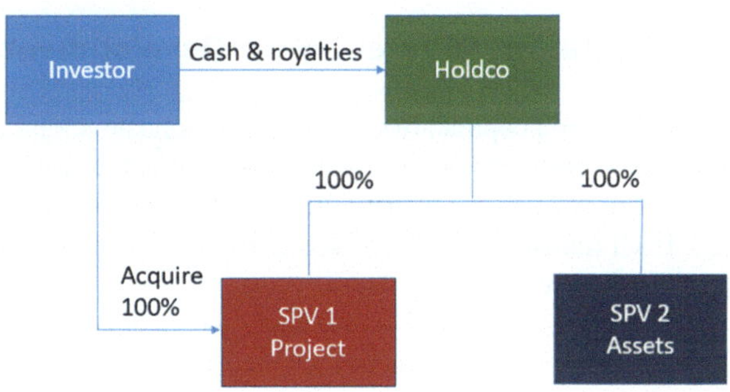

Fig. 4.18 Transaction structure—disinvestment finance (Source: The author)

Notes

- A key driver is the valuation of the assets to be sold by an independent valuator.
- The consideration for the disinvestment can be cash or cash and a royalty.

Example

- *Coal mining company, in financial distress*
- *Requires $3 million to settle creditors*
- *Investor: purchases project from coal mining company*
- *Consideration: $1 million plus $1/ton up to a maximum cash consideration of $3 million*

Mining capital type	Mine development stages				
	Exploration	Feasibility	Mine design	Construction	Operation
Equity finance	X	X			
Farm-in finance	X	X	X		
Listing	X	X	X	X	X
Reverse listing	X	X	X	X	X
Project finance			X	X	X
Development finance			X	X	X
Corporate bond finance			X	X	X
Streaming finance			X	X	X
Royalty finance			X	X	X
Offtake finance			X	X	X
Insurance finance				X	X
Equipment finance				X	X
Disinvestment finance	X	X			X

Fig. 4.19 Mining capital for different mine development phases (Source: The author based on Mining Journal, Global Mining Finance [5])

4.3 Mining Capital for Different Mine Development Phases

The different types of mining capital are specific to the mine development stage. The investor targeting an investment in an exploration project is different from the investor investing in a mine under construction or the investor investing into an operational mine. Each mining capital type has a different mandate and is targeting projects at specific mine development stages.

The matrix below in Fig. 4.19 outlines the mining capital types applicable to the different mine development stages. As can be seen the exploration and feasibility stages of a mining project have fewer financing possibilities than the phases of mine design, construction and operation. Operational mines have most mining capital types available to them, subject to the mining business case being commercially viable.

4.4 Combining Mining Capital Types

In the current mining investment climate, the winning formula is often a combination of different types of mining capital. This is the case when one funding method alone is not enough to provide all the CAPEX to develop / finance the mining business case. Multiple combinations exist, and the following combinations are more commonly used for mining investment.

4.4.1 Combination: Project Finance and Equity Finance

In this instance, the bank provides project finance to the SPV. Completion guarantees are provided by the Holdco. At the same time, a mining fund invests

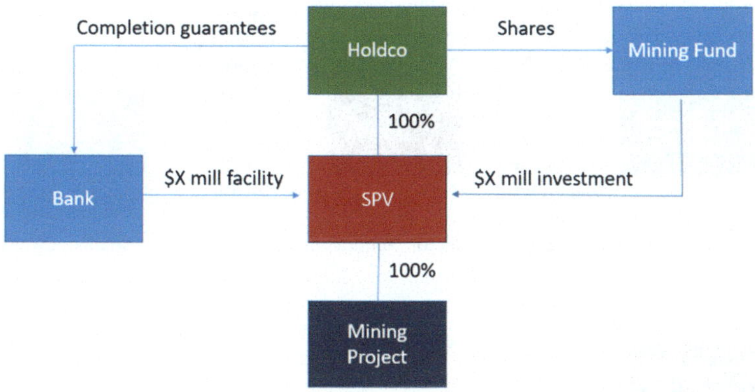

Fig. 4.20 Transaction structure—project finance and equity finance (Source: The author)

capital into the SPV, for shares in Holdco (Fig. 4.20). Banks prefer a strong equity partner, who will be able to share the financial burden in times of distress, making this combination an attractive option for financing mining projects.

4.4.2 Combination: Contractor Vendor Finance and Trader Finance

This financing route can be explored for mining projects when there is no traditional debt or equity finance available. The commodity trader provides offtake, streaming—or royalty finance. The collateral is mineral product. The EPC and O&M contractor is appointed for an extended period by the SPV to provide mining and/or processing CAPEX and services via contractor vendor financing. Capital is redeemed via operating costs which include larger margins or a profit share for the contractor (Fig. 4.21).

4.4.3 Combination: Insurance Wrapped Finance and Trader Finance

This route uses a combination of trader finance (offtake, streaming, royalty) and insurance wrapped financing. This type of finance is applicable for smaller mining projects in frontier countries, where the transaction is too small for a large commercial bank to participate.

The commodity trader provides offtake, streaming—or royalty finance. The collateral is mineral product. A regional bank funds the mining equipment and mining services, enabling goods and services to be provided to the SPV, subject to an insurance company insuring the bank. The SPV pays the mining services and equipment suppliers, who in turn, repay the bank (Fig. 4.22).

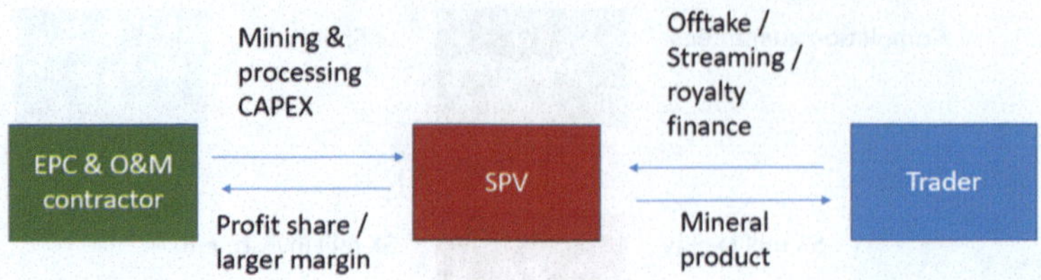

Fig. 4.21 Transaction structure—contractor vendor finance and trader finance (Source: The author)

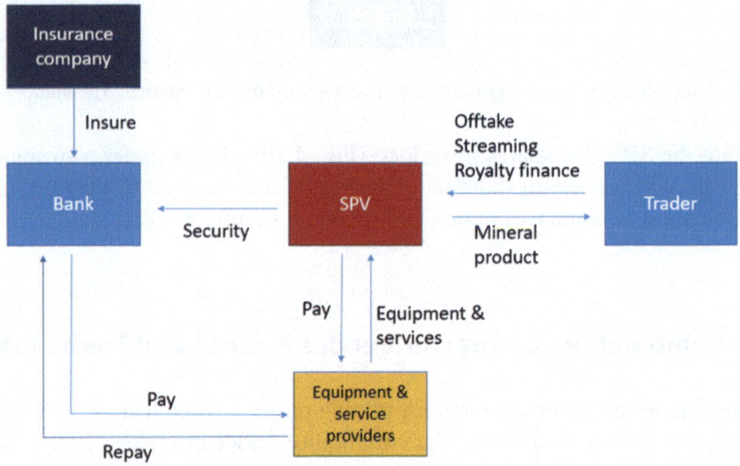

Fig. 4.22 Transaction structure—insurance wrapped debt finance and trader finance (Source: The author)

4.5 Mining Capital Due Diligence

The due diligence study is a key event on the road to securing mining capital. The prospective investor needs to undertake a thorough due diligence study on the mining investment proposal. Likewise, the mine developer needs to undertake a due diligence study on the investor.

4.5.1 Mining Investor Due Diligence

Once the targeted investor has the business plan and data, the due diligence study commences. This process is lengthy and includes clarification meetings and multiple conference calls to ensure that the investor has a full understanding of the investment proposition. Mining investors often appoint independent experts to conduct the due diligence process. These experts, usually geologists, mining engineers, metallurgists, infrastructure engineers and lawyers do not have any shareholding in the project or mine nor previous relationship with the mining project, and the aim is to be completely objective when assessing the viability and risks of the investment opportunity. A positive outcome usually means investment into the project, whereas a negative outcome will lead to no investment. The key questions that must be answered during the investor due diligence process are outlined in Table 4.4.

4.5.2 Site Visit

It is vital to do a mine site visit as part of the due diligence exercise. A site visit demonstrates to the prospective investor whether the mine developer's technical documentation (feasibility study, EIA, etc.) reflects the realities of the planned/existing mine site.

The informed investor should never invest in a mining project that he/she has not visited.

Even if the project is no more than a piece of land for a planned greenfields project, it is important to go to site and walk it, consider the setting and visualize a mining operation, as part of the due diligence exercise.

Mining investments have gone wrong simply because the mining investor has not been to site and missed that the site-specific conditions do not match the corresponding paperwork.

The site visit requires the mine developer to have the relevant large-scale plans of the mine infrastructure, mining block plan and logistics path at hand as a reference point for discussions on site.

Key observations for the mine site visit are outlined in Table 4.5.

4.5.3 Mine Developer's Due Diligence on the Investor

The mine developer needs to ensure that the mining funder is a genuine investor with capital or capital commitments to enable the transaction to happen after positive conclusion of the due diligence study. The following questions in Table 4.6 need to be used by the mine developer to determine that the ability finance.

4.6 Mining Capital Term Sheets

In this chapter, various mining capital term sheets are outlined for different types of mining projects, project stages and commodities (Fig. 4.23). The term sheet sets out the terms under which the mining investment will be made and serves as a template to develop the detailed legal agreement.

Once the parties involved reach an agreement on the details laid out in the term sheet, a binding agreement is drawn up and signed, prior to release of the funds into the project.

Table 4.4 Mining due diligence—key questions per discipline

Discipline	Key question
Legal issues	• Mineral and surface right ownership • Prospecting permit/mineral lease/mine authorization/mining right terms and conditions • Option/JV terms and conditions • Mine health and safety regulations and compliance
Geology	• Mineral abundance and distribution of occurrences in area of interest • Potential of the metallogenic province to host large and economic viable deposits • Ability to apply geological assessment techniques *Deposit scale:* • Local geological relationships • Deposit type • Mineralization characteristics • Size of deposit • Grade • Continuity • Product quality (chemical and physical) • Ability to apply exploration techniques • Resource/reserve classification • Results of geotechnical studies • Results of geohydrological studies—groundwater level and pump capacity • Depth of weathering—influence of ore composition and extraction
Exploration	• Drill spacing • Drilling method • Number of holes and metres drilled • Number of trenches • Mapping scale • Surveying method and accuracy • Other techniques used—geophysics and soil sampling • Analytical procedure used • Check sampling
Mining	• Probable mining method • Recoveries • Production capacity • Mining contractor's scope of work, quotations received and selection • CAPEX • OPEX
Processing	• Beneficiation testwork results • Probable processing techniques • Recoveries • Production capacity • Range of products • Processing contractor's scope of work and quotations CAPEX • CAPEX • OPEX
Infrastructure	• Mine infrastructure layout/mine block plan • Tailings dam design • Mine access • Power availability

(continued)

Table 4.4 (continued)

Discipline	Key question
	• Water availability • Topography • Climate • Availability of fabrication and maintenance services in area • CAPEX • OPEX
Marketing	• Mineral product specification • Distance from market • Current market conditions and demand/price outlook • Offtaker company details • Mineral sales agreement
Logistics	• Logistics path—mine to market • Distance to market • Method of transport • Logistics contractor • OPEX
Environmental and social issues	• Environmental sensitivity • Social impact of mining • Potential social benefits to local community • Rehabilitation bond amount
Human resources	• Who is the team behind the project and what are their skills and experiences? • Manpower required • Availability of skilled local labour
Financial	• Economic evaluation of project in terms of IRR (Internal rate of Return) and NPV (Net Present Value) taking the following into consideration: – Production data – Commodity prices – Revenue – CAPEX – OPEX – Tax Discount rates • Sensitivity analysis—change of key parameters by +10% and -10% and effect on NPV and IRR • Expenditure to date on project

Source: Authors table, based on Smith, L.D. Checklist for economic evaluations of mineral projects [54]

The mine developer can use the generic term sheets in this chapter as a reference for transactions for the different types of mining capital and apply and modify them to suit his/her specific project financing.

Note: For the term sheets provided in this chapter, 'X' represents different values and can be replaced by specific numbers for the specific mining project and related mining capital.

Table 4.5 Mine site visit—key observations

Discipline	Key observations during site visit
Mine site accesses	• Conditions of road • Ease of access to site • Terrain • Distance from nearest town
Exploration	• Location of drill holes • Presence of core
Mining	• Location of box cut/shaft • Mine production plan: direction, start and end
Mine infrastructure	• Location of mine infrastructure • Review of mine block plan
Mineral processing	• Location of process plant
Mine water	• Source of water • Extraction/delivery point • Availability of water on site
Mine power	• Power lines in the vicinity/distance
Engineering	• The availability of engineering services (workshops, suppliers) in the vicinity of the mine site
Human resources	• Where do the people live that will be employed by the mine? • A visit to the mine camp
Logistics	• Logistics path from the mine site for mineral commodity • Can the site accommodate the amount of planned trucks to enter the site?
Community	• Proximity of houses/dwellings • Prevailing wind direction • The effect of noise, dust and blasting on the community • Nearest town: a visit to the nearest town to get an understanding of the effect that mining will have on the community
Environmental	• A visit to the environmentally sensitive areas
Security	• Points of weakness • Security requirements

Source: Authors table

Table 4.6 Mine developer questions to investor

Category	Mine developers questions to investor
Investment mandate of the funder	• What type of mining project is the investor interested in investing in? – Commodity preference – Project stage • Amount of funds available per investment, minimum and maximum investment
Availability of capital	• Is there capital available for the type of investment considered? • When does the capital become available? • What amount of capital does the investor want to invest in the mining project?
Proof of funds	• Can the investor demonstrate proof of funds? This is important for less known finance sources. This involves a letter from the bank, confirming proof of funds.
Source of funds	• Where is the mining finance originating from?
Terms of funding	• What are the terms of funding? • For debt financing: interest rate, duration of loan, securities required • For equity financing: shareholding required in the project • Can the investor produce a draft term sheet from consideration by the mine developer?
Duration of financing	• How long will it take from due diligence to financial closure?
Team, key individuals behind the capital	• Name, qualification and expertise of the investment team members • Track record of investing in mining projects

Source: Authors table

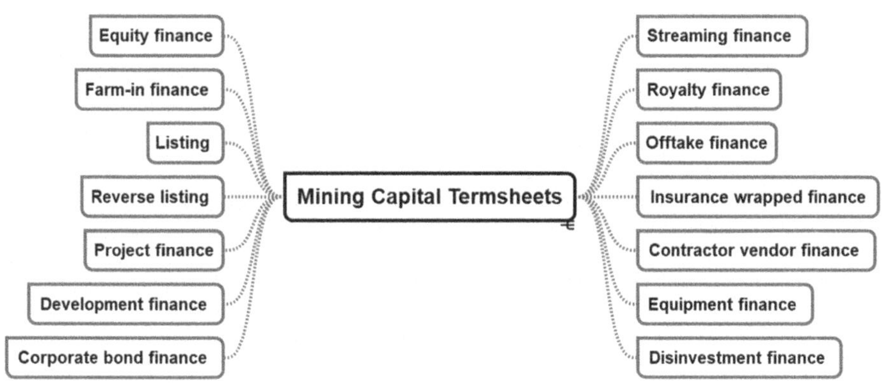

Fig. 4.23 Mining capital term sheets (Source: The author)

4.6.1 Term Sheet: Equity Finance

Equity finance involves an investor investing capital in exchange for shares in a project/company (Table 4.7).

Table 4.7 Term sheet: equity finance

Investor	• Mining fund
Investment	• $X million
Shares	• X new fully paid ordinary shares
Share price	• $X/share
Terms	• The investment must be used for financing the ongoing investment programme for the exploration, development or expansion of the project • Mining Fund will have the right to nominate one person as a director of company
Conditions precedent	• No breach of warranties • Performance by SPV • Warranties given by the SPV remain true and correct • Compliance by the SPV with the social, environmental, health and safety policies that have been approved • Mining fund will not be the single largest holder of shares in the company
Termination rights	• Mining fund may terminate by way of notice in the event of: – The Holdco or SPV being insolvent – A material breach of the warranties – If the number of ordinary shares allocated to mining fund would be less than 5% of the shares in the SPV.

Source: Authors table based on experience of raising equity projects for diverse mining projects and companies

4.6.2 Term Sheet: Farm-in Finance

Farm-in finance involves an investor, a strategic partner, investing into the mining project in phases and acquiring a staged increase in shares in the project with stages investment (Table 4.8).

Table 4.8 Term sheet: farm-in finance

Developer	• Holdco: owns 100% of SPV • SPV: owns the prospecting permit
Investor	• A mining company that wishes to acquire shares in the SPV • Willing to invest in the SPV staged phases and Farm-in to the SPV
Investment requirements	• Phase 1: $X million to complete the concept study • Phase 2: $X million to complete the feasibility study • Phase 3: $X million to construct the mine
Conditions precedent	• SPV allows the investor to conduct a legal and technical due diligence • The Investor has the exclusive right to acquire shares in the SPV
Payments	• Phase 1: The investor pays the SPV $X million as part of acquiring 15% of the prospecting permit, which will be used to complete the concept study. • Phase 2: Investor pays the SPV $X million as part of acquiring 50% of the prospecting permit to pay for the feasibility study. • Phase 3: SPV secures project finance to finance the project.
Shares	• Phase 1: Upon the conclusion of the concept study, Holdco issues investors 15% of the ordinary shares of SPV. • Phase 2: Upon the feasibility study, Holdco issues investors a further 35% of the ordinary shares of SPV. The investor will have acquired 50% of the issued shares by investing $X million in the SPV.
Timelines	• Payment 1 must be executed within 10 days of signing the Farm-in agreement. • Payment 2 must be executed within 2 years of signing the Farm-in agreement.
Option:	• Holdco grants the investor the option to acquire the reminder of shares of the SPV at market-related prices, determined by independent valuator.
Warrants	• Holdco warrants the investor that the prospecting licence: – is correctly registered – in good standing – has not been offered – has no environmental claims

Source: Authors table based on experience of concluding farm-in financing / strategic financing for mining projects

4.6.3 Term Sheet: Listing

Listing involves securing mining capital via listing the mining company on a public stock exchange. The terms specific to each stock exchange, but should cover the following as outlined in Table 4.9:

Table 4.9 Term sheet: listing

Company to be listed	• Company PLC
Valuation	• Independent valuation determined the project value to be $X million
Investment	• $X million
Shares	• X million new fully paid ordinary shares
Share price	• $X/share
Terms	• The investment must be used for financing the project development stages
Shares structure	• Issued shares: X shares • Holdco shareholders (developer): X shares • Public shareholders: X shares
Holdco shareholders	• Issued X shares at $X/share for vending their mining project into PLC • This equates to X% of the company
Public shareholders:	• To pay $X/share for y shares
Share price	• $X/share
Shareholding	• Holdco shareholders X% • Market X% • Institutional investors X% • Retail investors X%
Advisor	• Reputable company X appointed as corporate financial advisor
Independent mining consultant	• Independent mining consultant to conduct due diligence report on project and provide valuation
Legal firm	• To conduct due diligence on company secretarial documentation and agreements
Documentation	• Prospecting Licence • Competent Persons Report on the project(s) • Holdco company secretarial documentation • Audited financials of Holdco • Holdco team member's CVs • Proof of expenses • Project plan • Information Memorandum on mining investment, outlining business case
Escrow period	• Holdco shareholders may not sell their shares for X years
Directors	• X executive directors nominated from Holdco • X non-executive directors
Timing	• X months
Fees	• Holdco shareholders to pay $ X for listing process
Annual stock exchange fees	• $ X
Reporting	• PLC to publish: – Announcement after drilling campaign – Quarterly reports – Annual reports
Shareholders meeting	• To be held annually

Source: Authors table based on experience of listing a junior mining company

4.6.4 Term Sheet: Reverse Listing

Reverse listing entails the mine developer vending in the mining project, which is valued, into listed company in exchange for capital to develop the project (Table 4.10).

Table 4.10 Term sheet: reverse listing

Project	• Mining project X
Valuation	• Independent valuation determined the project value to be $X million
Shareholders	• Holdco, X% owner of the SPV
Project company	• SPV
Investment required	• $X million
Company listed	• PLC • Cash reserves: $X million
Market cap before project vend in	• $X million
Shares structure before project vend in	• Issued shares: X million shares issued • Share price: $X/share
Shares issued to Holdco shareholders	• Shares issued: X million shares issued • Share price: $X/share
Market cap of PLC post vend in	• $ X million
Commitment by PLC	• Use of PLC cash reserves $X million to develop project. • Raise capital on stock exchange to develop project

Source: Authors table based on experience of reverse listing a project into a listed junior mining company

4.6.5 Term Sheet: Project Finance

Project finance entails the mine developer lending capital from a commercial bank to finance the project. Term sheets are specific to the commercial bank providing project finance, but a representative term sheet structure is provided in Table 4.11.

Table 4.11 Term sheet: project finance

Lender	• Commercial bank
Borrower	• SPV
Shareholders	• Holdco X% and local partner X%
Sponsor	• Holdco
Obligors	• SPV, local partner and the sponsor jointly
Facilities	• Senior debt facility and Standby debt facility
Senior debt facility	• $X million to fund the capital expenditure of the project
Standby debt facility	• $X million to fund operating cost overruns or working capital shortfalls
Standby equity facility	• The sponsor to cede a standby equity account containing no less than $Y million to the lender to fund any capital and/or operating cost overruns or working capital shortfalls
Ranking	• Senior
Final maturity date	• Month, year
Financial close	• The date on which the lender notifies the borrower that the conditions precedent to financial close have been fulfilled satisfactory to the lender
Arranging and underwriting fee	• X% of the facility amount to be paid in instalments on credit approval, signature of the finance documents and financial close
Availability period	• X years from financial close
Commitment fees	• X% per annum
Interest rate	• The sum of base rate and the applicable margin
Base rate	• 3-month applicable rate to commercial bank
Default interest	• The interest rate plus X%
Interest payments	• Quarterly
Principal repayments	• Capital is to be repaid quarterly commencing x months from financial close
Voluntary prepayment penalties	• Voluntary prepayments allowed subject to a x% penalty
Mandatory prepayments	The lender will have the right to request prepayment of the facilities in certain instances: • a change in control • the sale of all or substantially all the assets • receipt of material insurance proceeds
Project accounts	• Proceeds account • Debt service reserve account • Distributions account
Proceeds account	• All revenues received by the borrower will pass through the proceeds account. • Ceded as security

(continued)

Table 4.11 (continued)

<table>
<tr><td>Debt service reserve account</td><td>• To be maintained at an amount equal to the next 6 months' payments of capital and interest due.
• Pledged and ceded as security</td></tr>
<tr><td>Cash sweep</td><td>• 50% of all excess cash in the proceeds account to be applied in prepayment of the standby debt facility</td></tr>
<tr><td>Payment waterfall</td><td>• All payments shall be subject to a conventional payment waterfall structure, as follows, in priority:
1. Taxes and royalties
2. Fees and costs due
3. Operating and capital costs
4. Interest on the senior debt facility
5. Capital on the senior debt facility
6. Interest on the standby debt facility
7. Capital on the standby debt facility</td></tr>
<tr><td>Security</td><td>• Full security over all the assets and mining licence of the borrower
• Completion guarantees from the sponsor
• Pledge of shares</td></tr>
<tr><td>Completion</td><td>• Completion: passing of technical, operational and economic completion tests developed in conjunction between the borrower, lender and the independent technical consultant appointed by the lender</td></tr>
<tr><td>Independent technical consultant</td><td>• To be appointed by the lender
• To be acceptable to the borrower
• Scope of work:
– Review of technical information for the project
– Monitoring of construction progress
– Verifying the cost to reach completion and performing completion tests</td></tr>
<tr><td>Financial covenants</td><td>The following financial covenants shall apply and shall be measured semi-annually on both a historic and forward-looking basis:
<table>
<tr><th>Covenant</th><th></th><th>Default level</th><th>Plan</th></tr>
<tr><td>Debt service cover ratio</td><td>No less than</td><td>1.4</td><td>1.6</td></tr>
<tr><td>Project life cover ratio</td><td>No less than</td><td>1.6</td><td>1.8</td></tr>
<tr><td>Reserve tail ratio</td><td>No less than</td><td>30%</td><td>30%</td></tr>
</table></td></tr>
<tr><td>Distribution conditions</td><td>The borrower shall not pay, make or declare any distribution to shareholders unless:
• Project completion is achieved
• First principal repayment is made
• The financial covenants are above the distribution levels as at the cost recent calculation date.</td></tr>
<tr><td>Conditions precedent:</td><td>• Satisfactory conclusion of due diligence
• The sponsor's equity commitment to have been spent on CAPEX.
• Satisfaction with all material contracts,
• All licences in place
• Satisfaction with the project in respect of the equator principles.</td></tr>
</table>

Source: Authors table based on experience of securing project finance for diverse mining project

4.6.6 Term Sheet: Development Finance

Development finance entails mining capital obtained from a development finance institutions (DFI), which provide debt finance for mining projects with developmental agenda for the host country/borrower (Table 4.12).

Table 4.12 Term sheet: development finance

Lender	• Development finance institution
Borrower	• SPV
Guarantor	• Holdco
Amount of facility	• $X million
Interest rate	• X% pa
Tenor	• The tenor of the loans is X months
Loan documents	• The loan agreement includes: – conditions to funding – representations and warranties – affirmative and negative covenants – information covenants – events of default – indemnity and boilerplate provisions – voting rights – security sharing – guarantees – share retention – subordination
Security	• Security will be first ranking security over: – All the SPVs assets – The SPVs bank account – Intercompany loans amongst SPV and the guarantors
Transaction documents	• Financing documents • Project documents • Letters of intent with respect to construction contracts and mining services agreement.
Legal/compliance	• SPV and Holdco to provide corporate authorization documents.
Miscellaneous	• Due diligence on the project • Project conditions: ore reserves report • Insurance: SPV must have adequate level of insurance in place and certification obtained from SPVs' insurers • Financial model: An agreed financial model, which includes a mine plan • Environmental and social conditions: – a completed social and environmental assessment – an environmental and social action plan – annual monitoring report – implementation of an agreed management system – appointment of a third-party consultant

(continued)

Table 4.12 (continued)

Default event	• A default arises for the SPV when one or more of the following key events is triggered: – Non-payment – Breach of representation or undertaking – Government expropriation or nationalization – Insolvency – Guarantors' liabilities exceeding $X million – Authorizations not maintained or restored within 30 days of a notice requiring restoration – Non-performance of project documents – All or a material part of the secured property is destroyed, lost or damaged beyond repair

Source: Authors table based on experience of financing of mining projects with DFIs

4.6.7 Term Sheet: Corporate Bond Finance

Corporate bond finance is debt finance facility with bonds as financial instrument, where Holdco issues bonds to the investor in exchange of mining capital. A generic term sheet is outlined in Table 4.13.

Table 4.13 Term sheet: corporate bond finance

Financial instrument	• Secured Redeemable Bonds
Corporation:	• Holdco
Holder	• Investor
Period	• X years
Interest	• X% pa
Value per bond	• $X
Value of issued bonds	• $X million
Trustee	• Acts as the administrator and paymaster
Terms	• The Corporation promises the holder to pay on X on presentation of the bond the sum of $ X plus interest at the office of the trustee • Bonds are issuable as fully registered bonds in denominations of $ • All bonds rank equally and with priority of preference • The bonds will not be redeemable on or before (date) • After (date), the bonds will be redeemable at the corporations sole option on X days' notice

Source: Authors table based on experience of financing a metals mining project with a corporate bond

4.6.8 Term Sheet: Streaming Finance

Streaming finance is mining capital provided by a streaming company to a mining company to construct a mining project, in exchange for securing metals, usually gold and silver at discount to market process (Table 4.14).

Table 4.14 Term sheet: streaming finance

Seller:	• Mining company (SPV)
Purchaser:	• Streaming company
Amount	• $X million
Stream agreement:	• Streaming company will provide mining company with a deposit of $X million to be used entirely towards the construction of the project. • Mining company agrees to deliver X% of the projected annual metals production from the project until X ounces have been delivered X% thereafter. • Streaming company purchase price per ounce will be X% of the London spot price for one ounce of refined gold at the time of delivery.
Obligations purchaser:	• Purchaser will provide the mining company with a $X million deposit. • The deposit will be provided over five instalments: – $X million on signing of the agreement – $X million on closing – X instalments of $X million
Obligation by company	• The company will use the proceeds towards the construction of the project. • Mining company agrees to deliver X% of the annual projected metal production from the project to the streamer for up to X ounces, and X% thereafter. • Streamer will pay the company in cash X% of the spot price for each delivered ounce.
Term	• The term of this agreement is X years • The term shall be extendable at either party's option for successive x year periods.
Covenants	• Conduct of Operations: All decisions regarding the development and operation of the project shall be made by seller in its sole discretion. • Seller and purchaser shall always maintain their corporate existence. • The seller shall not process other minerals through the facilities. • The seller shall maintain insurance for the project. • The purchaser has security interests in the assets of the project. • If the seller arranges senior financing that is to be secured by any senior financing party, the purchaser will enter into a subordination agreement with the relevant senior financing party. • The seller hereby agrees that it shall not sell unprocessed ore. • The seller shall complete a minimum x m drilling programme.

Source: Authors table based on experience of financing of gold projects with streaming finance and related market research

4.6.9 Term Sheet: Royalty Finance

Royalty finance is mining capital provided by a royalty company to a mining company to construct a metals mining project, usually gold or silver, in exchange for a royalty, which is a percentage of the gross value of recoverable minerals (Table 4.15).

Table 4.15 Term sheet: royalty finance

Owner	• The shareholder (SPV) of the project seeks to grant to the royalty holder a net smelter returns royalty
Investment	• $X million
Production	• The quantity of refined metal that is produced to the owner's account by a refinery during a month
Monthly average metal price	• Average London Bullion Market Association metal price
Net smelter returns	• Gross proceeds for such month from the sale of such product less allowable deductions for such month
Products	• All metal-bearing ores mined from the project and all concentrates and other mineral products, metals or minerals, which are derived therefrom
Net smelter return royalty	• In consideration of the payment of the investment which shall be paid by the royalty holder to the owner, the owner agrees to pay to the royalty holder the royalty in respect of products
Royalty	• X% of the gross value of recoverable minerals contained in such products
Term	• The entire term of the mining lease
covenant	• The royalty will be running with the project
Tailings and residue	• All tailing, residues, waste rock, spoiled leach materials, and other materials are subject to the obligation to pay the royalty
Purchase of royalty	• $X million
Warranties	• The owner warrants to and in favour of the royalty holder that it has the corporate company power, capacity and authority to execute, deliver and perform this agreement
Transfer rights of the royalty holder	• The royalty holder shall have the right to transfer its rights and obligations under this agreement to one or more lenders providing financing to the royalty holder
Owners discretion	• The owner will have complete discretion concerning the nature, timing and extent of all exploration, development and mining of the project

Source: Authors table based on related market research

4.6.10 Term Sheet: Offtake Finance

Offtake finance is mining capital provided by a bulk commodity trading company to a mining company to execute a mining project which is close to production. Offtake finance is the form of a prepayment to fast-track a project into production (Table 4.16).

Table 4.16 Term sheet: offtake finance

Seller	• Mining company SPV
Buyer	• Commodity trading company r
Product	• Commodity: X • Quantity: X tons per month • Quality: X • Sizing: X
Period	• X months
Start date	• Date X
Price	• Commodity indexed based price, less marketing fee of x%
Delivery point	• As defined by buyer
Prepayment	• 100% prepayment to be made by buyer, X business days prior to the commencement of the month that mineral product will be delivered to delivery point.

Source: Authors table based on offtake finance transaction for coal projects

4.6.11 Term Sheet: Insurance Wrapped Finance

Insurance backed finance is a method of financing mining equipment and mining services, which form part of the capitalization plan of a mining project. The presence of an insurer allows a debt facility provider to provide a loan for the benefit of the project, knowing that it is insured in an event of default (Table 4.17).

Table 4.17 Term sheet: insurance wrapped finance

Borrower	• SPV: provides business case to capitalise mining project to the value of $X million • The bulk of the capital is for mining equipment
Insurance company	• International insurer approves the business case and insures the bank against a default of the business case
Bank	• A commercial bank provides a debt facility to recapitalise mine • The mine equipment supplier and mine service providers are paid by the bank for equipment and services to be rendered to recapitalise the mine
Equipment suppliers	• Equipment supplier X provides mining equipment:
Requirements by bank and insurer	• A business case: • Mining licence is intact • Mineral resources and reserves • A mine plan • A mine development and operations team • A market for the mineral product
Facilities:	• Value of mining equipment: $X million
Interest rate	• X% pa
Term	• X years
Security	• Mining licence of SPV

Source: Authors table based on insurance backed debt financing structures for mining equipment in political high-risk countries

4.6.12 Term Sheet: Contractor Vendor Finance

Contract vendor finance is finance/mining capital provided by large, financially strong contractors of a mining project, who capitalize the required equipment for a mining project (mining fleet, processing plants, power plants) in exchange of long-term contracts, over which the capital is redeemed (Table 4.18).

Table 4.18 Term sheet: contractor vendor finance

Funder	• Mining contracting company
Borrower	• SPV
Contract type	• Build + Own + Operate (BOO)
Scope of work	• Mining contracting company to provide: • Mining equipment • Mineral processing equipment • Site establishment fees • Operational management team • Required production rate: X Mtpa • Production rate: X tph • Mineral product: quantity and quality
Requirements by Borrower	• Borrower to ensure: – Mining licence is intact – Geological model is available – Market for product guaranteed with short-, medium- and long-term contracts
Facilities:	• Value of mining equipment: $X million • Site establishment fee: $X million
Commercial rates	• SPV to pay contractor: – Fixed fee: $X per month – Variable fee: $X/ton
Term	• X years
Security	• Mining licence

Source: Authors table based on experience of financing mining projects with contractor vendor finance

4.6.13 Term Sheet: Equipment Finance

Equipment finance is mining capital that is provided by commercial banks to the mining and or processing contractor as a debt facility to purchase equipment to execute the contract on a mining project (Table 4.19).

Table 4.19 Term sheet: equipment finance

Lender	• Commercial bank
Borrower	• Contractor
Type of Agreement:	• Instalment sale agreement and credit facility
Description of goods	• Mining or processing equipment X
Cash value	• $X million
VAT Amount	• $X million
Total cash value	• $X million
Interest	• X%
$ value interest	• $X million
Total amount repayable	• $X million
Suretyship	• Shareholders of contractor
Number of instalments	• X
Amount of each instalment	• $X

Source: Authors table based on experience of financing mining equipment for mining contractors

4.6.14 Term Sheet: Disinvestment Finance

Disinvestment finance is mining capital raised by selling a non-core mining asset. Cash raised in the sale of the asset is used for developing other mining projects (Table 4.20).

Table 4.20 Term sheet: disinvestment finance

Seller	• Holdco
Investor	• Mining company
Asset	• A mining project that does not form of Holdco's business strategy
Transaction	• Investor will acquire X% of the asset, including and any surface rights, servitudes or road access held by Holdco which are required to develop the asset
Consideration	• A cash payment of $X million plus a royalty of $X per tonne produced until maximum cash consideration of $X million has been paid to Holdco
Warrants	• Holdco warrants that it shall cause no harm or liability to the mineral or surface rights
Condition precedent	• Completion of a due diligence process by the investor, which shall include drilling further holes on the mineral rights included in the transaction to expedite the mining and extraction process
Approval	• This offer is subject to the approval of the transaction by the boards of the investor and Holdco

Source: Authors table based on experience of raising mining capital by selling non-core mining assets

4.7 The Process of Raising Mining Capital

Raising capital for mining projects is a step-by-step process. The target is to raise the required capital for the mining project, considering the development stage of the project, the commodity, size and location. The process follows ten steps, from the funding scope to investment (Fig. 4.24).

4.7.1 Step 1: Mining Business Case

The process of raising mining capital starts with developing a solid mining business case from the reserve to the customer, depicted in Fig. 4.25 and referred to in detail in Sect. 2.3.

4.7.2 Step 2: Funding Scope

Once the mine developer has developed the mining business case, the exact funding scope needs to be determined and with the following questions to be answered (Table 4.21):

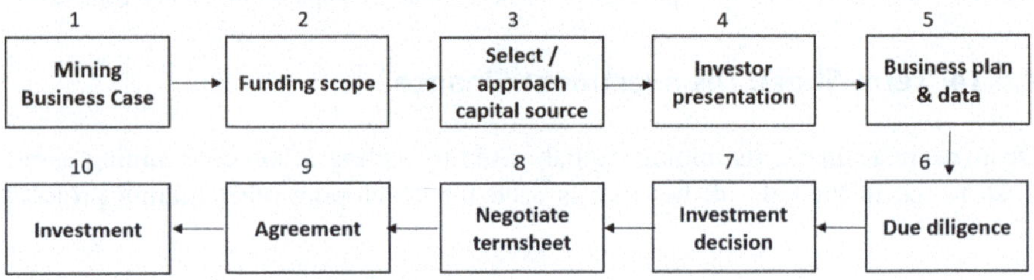

Fig. 4.24 The process of raising mining capital (Source: The author)

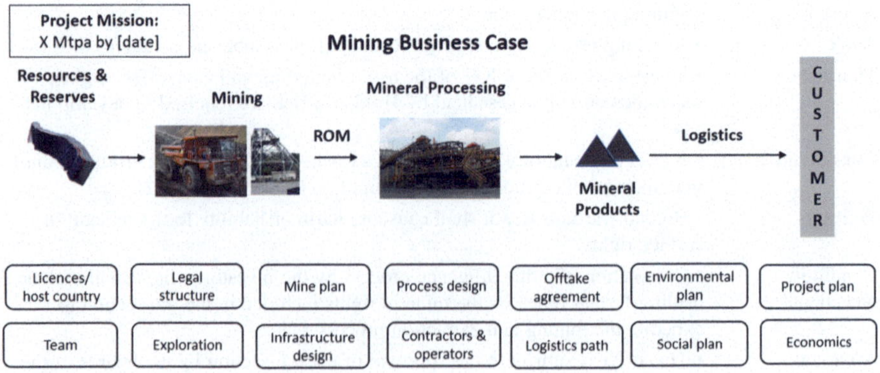

Fig. 4.25 Mining business case for raising capital (Source: The author)

Table 4.21 Mining project funding scope questions

Funding scope question	Information required
How much CAPEX is required?	• $X million • Required by when • Once off funding or funding in stages
What is the purpose of the CAPEX?	Options: • To develop the feasibility study • To conduct exploration • To commence with the development of the mine • To purchase mining equipment • To recapitalise the mine
What is the amount of capital spent to date?	A list of expenses incurred to date for: • Exploration • Licensing • Project acquisition • Mine development • Equipment • Acquisition • Business development
What is the offering to investors?	• Shareholding in the project/company • Interest and capital repayment • Royalties • Mineral product

Source: Authors table

4.7.3 Step 3: Select Mining Capital

Next, the mine developer needs to select the type of mining capital (Fig. 4.26) appropriate to the mining business case. The types of mining capital applicable to the project stage and commodity are discussed Sect. 4.2.

Once the mining capital type has been determined, it has to be sourced. An online search of key words will lead to various mining capital type providers directly or advisory firms working with mining capital sources. The search should be as follows (Table 4.22):

After the mining capital source has been identified, it is time to formulate an approach for financing. Below typical enquiry for mining capital:

* * *

Dear XXX

GOLD PROJECT SEEKING CAPITAL FOR DEVELOPMENT

We, the developers of the ABC Gold Project are seeking capital for the development of the project. The projects summary is provided below:

- *Advanced gold project in xxx mining province, with significant silver credits*
- *Location: xxx*
- *Fully permitted*
- *2-phased development approach*

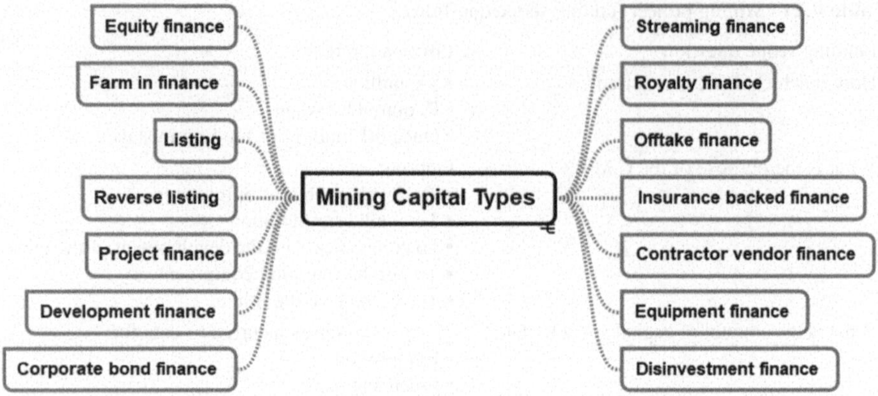

Fig. 4.26 Mining capital types (Source: The author)

Table 4.22 Identifying the mining capital sources

#	Main keyword	Example
1	Project stage	*Mine development project*
2	Commodity	*Gold, silver*
3	Mining capital type	*Streaming finance*

Source: Authors table

  - *Phase 1: Development of small profitable underground gold vein mining operation, which will produce 178 kg/annum gold and 111,000 kg/annum of silver*
  - *Phase 2 and 3: Ongoing exploration and feasibility to develop an opencast operation*
- *Feasibility study completed and market secured for Phase 1*
- *18 months to cash flow*
- *Capital required for Phase 1: $ 11 million*
- *Phase 1 economics: IRR: 26%, NPV (10%) $7.6 million, 3-year payback period*

We are enquiring about project finance or a corporate bond.

We have prepared a business plan and all the supporting data, which is available on request and the signing of an NDA. Please see attached the investor presentation.

We look forward to discussing the project in more detail with yourselves.

Sincerely,

Director

XXX Gold Mining Ltd

* * *

4.7.4 Step 4: Investor Presentation

Step 4 is to forward the investor presentation to the targeted investor. The presentation needs to be clear, concise and attract interest in the opportunity. The details of the good investor presentation are outlined in Sect. 3.2. The investor presentation needs the following slides to cover the scope of the investment:

- Project Summary
- Location
- Ownership
- Licences
- Leadership
- Geology and exploration
- Resources and reserves
- Development strategy
- Mining
- Processing
- Mine infrastructure
- Logistics
- Market
- Development team
- Environmental and Social
- Economic analysis
  - Input parameters
  - CAPEX and funding requirements
  - OPEX
  - Economic indicators
- Implementation plan

The Investor presentation in Sect. 3.2.3 is an example and can be tailored to suit the individual project seeking mining capital (Fig. 4.27)

4.7.5 Step 5: Business Plan and Data

If the investor presentation has attracted interest by the prospective investor and fits the investment mandate, the investor will want the business plan and supporting data.

Non-disclosure and Non-circumvention Agreement

For the parties to exchange information a non-disclosure agreement and non-circumvention need to be signed. Both the investor and the project to developer are the owners of proprietary information. The developer has proprietary project information, and the funder may have a network of co-financiers;both need to be protected.

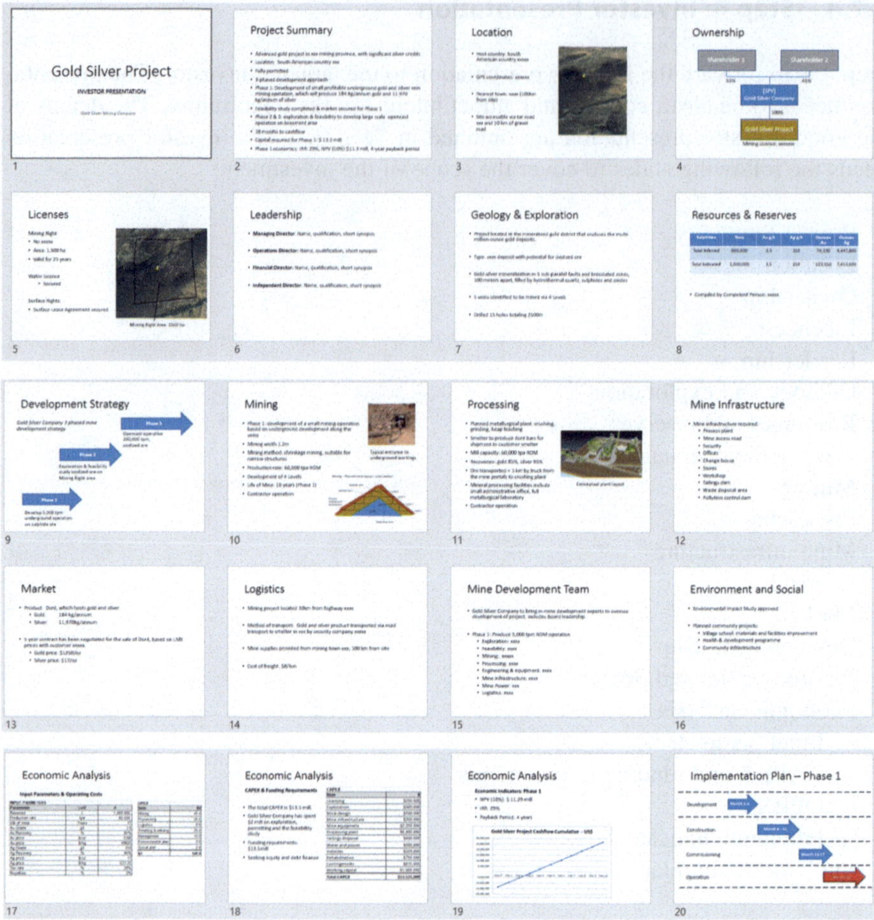

Fig. 4.27 Investor presentation example (Source: The author)

Business Plan

The business plan needs to be compiled according structure outlined in Sect. 3.3.2

1. Executive Summary
2. Vision and Mission
3. Location
4. Ownership Structure and Shareholders
5. History
6. Licences
7. Geology and Exploration
8. Resources and Reserves
9. Project Development Strategy
10. Mining
11. Mineral Processing

12. Mine Infrastructure
13. Market
14. Logistics
15. Human Resources
16. Contracting
17. Environmental Management Plan
18. Social and Labour Plan
19. Financial Analysis
20. Risk Management Plan

The business plan example of the Vanadium Titanium Iron Project in Sect. 3.3.3 can be tailored to suit the individual project seeking mining capital.

Data Room

The mining business plan requires supporting data. The data is technical, legal and financial and involves historical and new project data. The data room needs to be structured in a clear manner to enable the mining investor to conduct due diligence study. The data of mining projects is extensive, and it can be provided in the cloud. An example of a data room structure is provided below (Fig. 4.28):

4.7.6 Step 6: Due Diligence

Once the targeted investor has the mining business plan and data, the due diligence study commences as outlined in Sect. 4.5.

Key activities on summary:

- Review of data in the data room
- Clarification meetings
- Independent experts to conduct the due diligence process.
- Site visit

Equally important is the mine developer's due diligence of on the investor as outlined in Sect. 4.5.3.

Key pointers in summary:

- Investment mandate of the funder
- Availability of capital
- Proof of funds
- Source of funds
- Terms of funding
- Duration of financing
- Team, key individuals behind the capital

Fig. 4.28 Data room structure for due diligence (Source: The author)

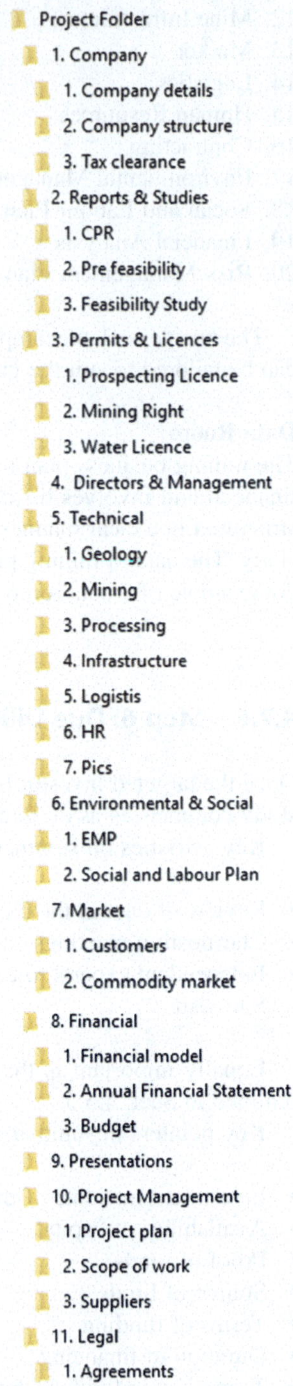

4.7.7 Step 7: Investment Decision

After the completion of the due diligence exercise, the investor will decide. The decision is usually made by a committee within the investor grouping, reviewing the outcome of the due diligence on the mining business plan, data room and site visit. Key questions the committee will deliberate on are:

- Does this investment opportunity meet our investment criteria and investment hurdle rates?
- Does this investment opportunity fit our profile?
- What are the risks, and can they be managed?

A 'yes' means the deal is on, the mine developer is informed about the positive outcome and a term sheet can be drafted for negotiation.

A 'no' means that the investment will not proceed, and the process comes to halt. The process of attracting an investor must restart.

4.7.8 Step 8: Negotiate Term Sheet

The next step is the negotiation of the term sheet. A term sheet needs to be negotiated between the mine developer and the investor. The term sheet forms the basis of the agreement. A solid term sheet makes for a good mining deal. Term sheets are structured individually on a project-by-project basis. Section 4.6 outlines proved term sheets that can serve as a reference for:

- Equity finance/share subscription
- Farm-in funding/strategic partnerships
- Listing
- Reverse listing
- Project finance
- Development finance
- Corporate bond finance
- Streaming finance
- Royalty finance
- Offtake finance
- Insurance backed finance
- Contractor vendor finance
- Equipment finance
- Disinvestment finance

Time and money can be saved when the term sheet is also depicted according to the transaction structures relevant to the mining capital types outlined in Sect. 4.2.

Once terms are agreed upon, the term sheet is signed by both the mine developer and the investor and the document is forwarded to their respective lawyers for conclusion of an agreement. Examples of term sheets for mining capital transactions are outlined in Sect. 5 Case Studies.

4.7.9 Step 9: Agreement

The term sheet is converted into an agreement. The agreement is an expansion of the term sheet and follows the same structure.

Both the mine developer and the investor appoint their respective attorneys to negotiate the finer details of the term sheet and draft an agreement, accordingly.

Note: the agreement needs to be simple and be drafted by lawyers who understand the mining industry and mining-related transactions.

It is recommended to use practical lawyers as early as possible in the process, ideally at the term sheet phase, in order to fast-track the process of drafting the agreement.

4.7.10 Step 10: Investment

This is the time for which all mine developers wait for: Mining capital funds flow from the investor's account into the mine developer's account to execute the mining business case.

Without this happening, nothing has happened, and all previous effort was worthless. So, wait for this event!

- For an equity transaction, a condition precedent is the issue of shares to the investor, which precedes the release of funds to the project.
- For a debt transaction, often a drawdown procedure is applied. The investment into the project follows in tranches. As project milestones are achieved, a new tranche of investment follows until the entire facility is consummated.
- For streaming, royalty or offtake finance, funds are released into the account of the mine developer in exchange for mineral commodities.
- For contractor vendor finance, the contractor commences with site establishment activities.
- For equipment finance, the equipment is funded and deployed to the mine site.

The mission of *mining capital* is accomplished!

4.8 The Human Dynamics of Mining Capital

Mining capital is as much technical and commercial as it is psychological. The task for the mine developer is to entice the mining capital provider to release money and transfer it into the mining project, ensuring that there will be an attractive rate of return for the invested capital to accommodate the risky environment of mining.

In this context, there are various human dynamics which the mine developer seeking capital or the investor seeking to invest capital in mining projects needs to be aware of (Fig. 4.29). These dynamics can be worth millions of dollars to a mine developer or investor who has fallen short in one of the issues mentioned below. They are worth discussing.

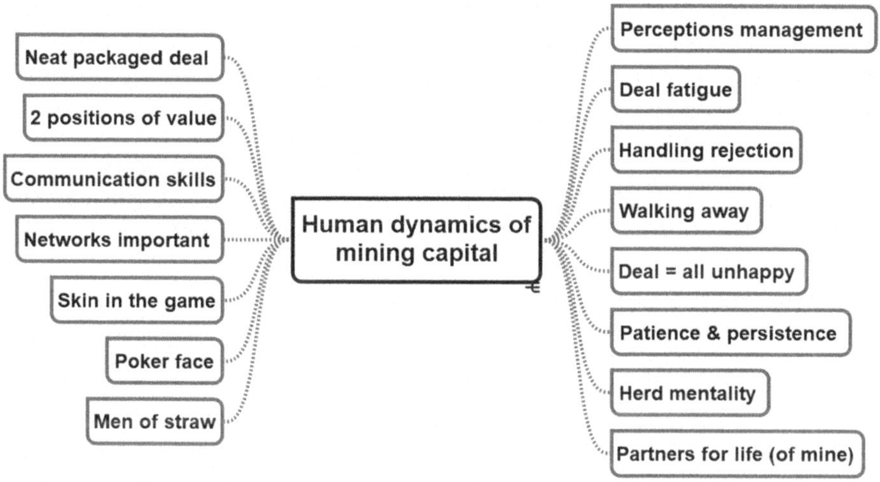

Fig. 4.29 Human dynamics of mining capital (Source: The author)

The Need for a Neat, Simple and Colourful Packaged Deal Proposal to Attract Interest

Mining deals and projects seeking capital need to be packaged in a manner that attracts investors. This must include pictures of the geology, mining equipment, processing plants, and mining operations in general. The owner of the money must be enticed to spend his money on a new mining venture. The impression of cash-generating mining assets or value appreciation needs to be generated and substantiated in the business case.

The Two Different Perceptions of Value

Section 2.7 covers valuation of mining projects. Sellers and promoters of mining projects have made a valuation of their projects before they attract investors. Section 2.7 covers valuation of mining projects. Mine developers have a perceived value of their investment proposition and are emotionally attached to their project.

The investor, however, sees an opportunity to acquire a stake in a mining project or operation with shortcomings and an opportunity to create shareholder value. The investor will try and downplay the value of the mining investment opportunity and will state that.

That is where the negotiations start. At two different value perceptions. The seller and the buyer will have to negotiate a value that both parties agree on, in order to move forward together. The seller does not want the buyer to leave, and the buyer does not want the seller to move on and look elsewhere.

Unless there are alternatives. Such negotiations are often accompanied by a fair share of bullying on the buyer's side and emotional outbreaks on the seller's side. This forms part of robust debate. If both parties eventually agree, they move forward to bring the project to the next stage as partners. If they do not like each other, they should part ways and look elsewhere for suitable project partners. Projects are doomed if a deal is concluded where the partners cannot work with one another.

Communication Skills

To an outsider, mining investments are often complicated, with multiple variables, technologically complex and involving risks with considerable investments. Many bankers keen to invest in mining projects are young and do not have mining experience. They are lost in mining terminology. The crux for the mine developer lies in communicating the mining investment opportunity in simple terms to people who are new to mining. Communication skills are key in bringing the investment opportunity across to the investor and motivating the investment into the mining project.

The Importance of a Network

Mining deals are often concluded by members within a network. Mining business people who know each other and share common experiences and education want to conclude a business transaction amongst themselves. How does one get into such a network? Mining investment seminars, studying mining courses, joining a mining business association, entering the mining industry as a consultant or supplier to the industry, acquiring a mining qualification are all good entry points into the industry, which is often difficult to get into. There is often the aspect of camaraderie that plays a role. Shared experiences in an industry that basically involves mining and breaking rocks to extract minerals in difficult environments forms bonds and networks for future potential business.

Skin in the Game

'Skin in the game' is a term often used by investor. It refers to the equity contribution of the mine developer. 'How much skin in the game do you have?' This means how much of your own money have you spent on the project or are you willing to spend further if the investor invests in your project. The developer is often perplexed by this question, approaching the investor precisely for that reason—to get money for the project. And here the investor is asking the developer to spend more money. The tactic behind it is that the investor wants to expose the weakness of the developer and secures himself a larger portion of the shareholding in the project. The investor wants to prove that the mine developer has little skin in the game, and hence it can take a bigger share of the project. Alternatively, the investor wants the mine developer to firmly demonstrate that he believes in the project. This firm commitment by the developer sends a signal to the investor that the developer will give it all to develop the project. How to approach this: Skin in the game needs to be:

- Money clearly spent on the project, auditable
- Ownership of property or assets in the project seeking capital

Skin in the game needs to be auditable and quantifiable in money terms, and hours spent on the project need to be logged and multiplied by an industry rate.

Poker Face

Mine developers spent time and effort to present their project credentials to a prospective investor. Once the presentation is delivered, there is often no reaction from the investor or a downplay of the project credentials. Behind the scenes, the investor may be excited about the opportunity, but to the developer he shows no emotions, i.e. a poker face. The purpose is to get maximum value for minimum cash investment. Watch out: a non-emotional reaction could also be a veiled poker face.

Men of Straw

'Men of straw' are business people, male and female, well dressed and give the impression of being an investor. However, when it comes to payment, no money flows, with much disappointment to the mine developer. They do not possess money, nor do they have access to money. They also intend to secure themselves an equity position and then use the data for their own roadshow.

Perception Management

A mining investment opportunity brought to a prospective investor needs to tick the key boxes of licences, assets in the ground, a market, a team, and a technical ability to produce mineral products. There are many variables that are outside the control of the developer: the geology, the site location, the market and the weather. Funding is about perception management. The funder needs to perceive that the mine development team can deliver on all aspects listed above and has a management plan in place to deal with the uncertainties, prior to releasing funds.

Deal Fatigue

If a packaged mining investment opportunity is promoted to several mining investors and brokers and roadshows and negotiations lead to no conclusive deal, deal fatigue can set in. Both the prospective investor and the mine developer become tired of the interaction and the project goes nowhere. At this point, a time out can be the best strategy. The project is temporarily shelved for a better day. The project may need to be packaged differently. Projects promoters should be aware of the concept of deal fatigue and focus on getting a deal over the line, if they can see a working partnership going forward.

Handling of Rejection

A mine developer promoting his investment proposal must be willing to handle rejection. Most mining investors will review and initially decline an investment opportunity and wait for a better deal to come around the table. But they do not come around the table too often and it is worth it to look again after adjustments have been made. The mine developer needs to ask the prospective investor the reasons for not investing and objectively identify shortcomings of his/ her proposal. Rejections are hard to handle but must be used as a guideline to improve the project.

Walking Away from a Deal

In some instances, the best outcome is to walk away from a mining deal and say 'no'. If the business case is lucrative on paper, but there is no physical match on the planned or prospective mine site, or the numbers are unrealistic, it is time to walk away from a mining deal. This can be by far the cheapest option as losses incurred by mining are substantial to any mine developer and mining capital provider.

A Workable Deal = Everyone Is Unhappy
A workable deal is a deal where both parties, the mine developer and the mining investor, have had to compromise and find the middle path to conclude the transaction. Often the parties feel that they have had to compromise too much. The mine developer believed that the investor has taken too much equity. The investor believes the mine developer has inflated the value of his contribution to the project thus far. Either way, to conclude the transaction both parties need to compromise and will potentially be 'unhappy'. That may be required to conclude the transaction.

Patience and Persistence
Mining transactions are large transactions, and capital providers take significant time in making investment decisions. It takes several years to implement a mining project, and there are several variables, some of which are in and some outside the control of the mine developer. A key to success is patience and persistence. Patience to develop a solid mining business case from orebody to market that attracts investment. Persistence to continue approaching targeted investors, despite encountering rejections.

Herd Mentality
The concept of herd mentality is very applicable to mining investments. Often, as soon as an investor decides to invest in a mining project, other investors follow suit, as they assume the due diligence has resulted in a positive outcome with good investment returns. They do not want to lose out on a good deal. Likewise, if one investor moves out of a mining investment, other investors will follow suit. The mine developer needs to be aware of the herd mentality of investors.

Partners for Life (Of Mine)
Once an investor invests large sums of capital in a mining project, believing in the team and the mining business case, the mine developer and investor become partners and enter a long-term relationship until the capital is repaid. For this relationship to flourish, regular communication of technical, operational, financial and legal matters pertaining to the mining business case are essential. The investor needs to know about both positive and negative events that may affect the mining business case, and transparency is essential. The mine developer needs to know that the investor wants the team to succeed. The mine developer and the investor become partners [55] (for life of mine).

4.9 Non-fundable Mining Deals: Examples and Action Plans

Some mining business cases cannot be financed. The principle reason is that the mining business case is flawed, and mine developers and promoters need to recognize that. During the boom time of commodity prices flawed mining business cases could be financed, these days the mining business case is subject to intense scrutiny and only the best mining business cases get financed. This is the main reason why mining capital is intrinsically linked to the mining business case.

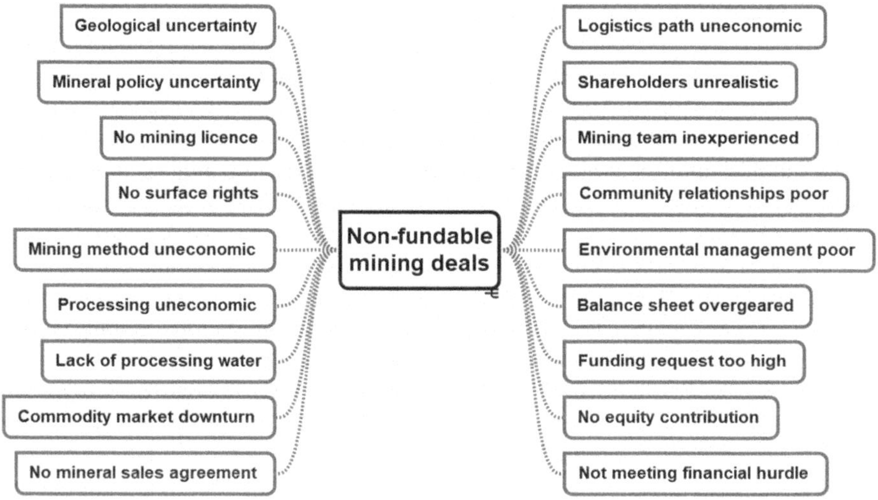

Fig. 4.30 Non-fundable mining deals (Source: The author)

It is far cheaper to not mine than to commence a mining operation with a flawed business case. If every ton that passes the conveyor belt incurs a loss, the financial damages quickly become excessive and out of control. Reasons why certain mining projects cannot be financed (Fig. 4.30) are provided below as well as in this chapter, with case studies and actions plans to address the shortcoming.

4.9.1 Geological Uncertainty

At the heart of any mining project is the geology with well explored, quantifiable economic resources and reserves, verified by a registered geologist who has experience in that geology, where the results are published in a CPR of an acknowledged institution. Following the BreX scandal [56], adequate systems and procedures have been put in place that govern geological certainty as large amounts of mining capital were lost. Geological uncertainty implies inadequate exploration data, and a resource reserve classification that does not warrant capitalizing a mining operation due to the high risk.

Example: A mine developer finances an underground mining operation with limited geological data, i.e. geological uncertainty. One year into production, the mining operation encounter of significant faults and dykes in the underground sections that required excessive waste mining, decreased the production performance, and resulted in unexpected increases in CAPEX and OPEX. After 2 years the mine is bankrupt, and the mining capital invested has been lost.

Action plan:

- Only develop a mine with proven and economic resources and reserves that have been signed off by a competent person, a registered and certified geologist in the applicable commodity, in a competent persons report.

4.9.2 Mineral Policy Uncertainty

Host governments formulate mineral policy rules around the extraction and beneficiation of mineral resources. Mineral policy governs primarily the procedure and process of awarding, prospecting and mining licences, local participation and empowerment and stakeholder engagement and royalties to the state. Mineral policy needs to simple, structured, consistent, and cognisant to the long lead times of mining projects and investment friendly to attract mining capital. The different mineral policies of different resource rich host countries compete with one another for mining capital. When there is mineral policy uncertainty due to regular changes to policy rules and regulations by governments yielding to political and social pressures and opportunism, mining capital moves on to other jurisdiction where there is more mineral policy certainty.

Example: The host government of a resource-rich country regularly changes the mineral resource policy to suit the current government agenda. Large mining companies decide to gradually sell their producing mines, stop new mining project development activities and gradually exit the host country. The once flourishing mining industry diminishes in size, and the host government fails to attract the required mining capital for new large and medium-scale mining projects.

Action plan:

- Persuade the host government, via the local chamber of mines and mining industry organizations that a consistent, dependable mineral policy is at the core of attracting global mining capital.
- Focus on attracting commodity-based offtake and royalty/streaming capital solutions to finance mining projects, as opposed to equity and debt financing.
- Sell/mothball mining project if no solution is found and move to more attractive mining jurisdictions, where investor friendly mineral policies will attract mining capital.

4.9.3 Licence Not Issued

The issuance of licences (exploration licence, mining licence and water licence) is not in the control of the mine developer. The host country does not issue the mining licence despite exploration and the feasibility study being concluded. The mine developer can try and expedite the issuance of the licence by engaging on regular intervals with the government authorities or bring a local liaison officer to interact with the authorities. The danger is that in some jurisdictions, individuals in government departments are seeking bribes to facilitate the licence, placing the mining company in a very difficult position.

Example: A major mining company develops a large-scale iron ore mining project. Large capital is invested into the exploration, feasibility and construction phase of the project. When the host government fails to issue the mining licence, the major mining company aborts the project and significant capital is lost.

Action plan:

- Only capitalize the mining project once the mining licence is issued.
- Regular engagement and communication with government officials to determine root cause of project
- Mothball project, if no solution is found.

4.9.4 Surface Rights Not Secured

The mining company fails to secure the surface purchase or lease agreement with the property right holder. Surface rights holders often have excessive price expectations when negotiating the sale of properties to mining companies. The project is delayed by years, and investors abort the investment process.

Example: A gold project cannot secure the surface rights required for the tailings dam from the property rights holder. The project is delayed by multiple years and interested investors move on.

Action plan:

- Downsize the project to avoid surface rights issues, if possible.
- Have the property value assessed by three agreed upon independent property valuators.
- Use of arbitrator to negotiate with surface rights holder.
- Include the surface rights holder into the mining business.
- Negotiate a royalty payment schedule for the surface rights holder.
- Only capitalize the mining project once the surface rights are secured.
- Mothball project, if no solution is found.

4.9.5 Mining Method Not Economic

The mine developer is overoptimistic and inexperienced about the mining costs and assumes the mine can be mined in-house or by contractors at economic rates. Despite resources and reserves, a good commodity market, proximity to market, the complex orebody results in too high mining costs, making the project uneconomic and unfinanceable.

Example: the mine developer overestimates the production performance of a narrow vein mining method, and the mining project economics do not work at lower production rates and higher unit mining costs. The project cannot be funded.

Action plan:

- Develop a detailed long, medium- and short-term mining schedule.
- Develop a scope of work and go out to tender to determine current mining costs by experienced mining contractors.
- If mining costs cannot be reduced, mothball project.

4.9.6 Mineral Processing: The Ore Cannot Be Processed Economically, Technology Risks

In certain complex orebodies, multiple metals are combined in the host hard rock and cannot be separated economically. Mineral processing costs become uneconomic.

In other instances, the process technology has proven itself on a pilot/laboratory scale but not on a commercial scale yet. No financing can be secured.

Example:

- *The ore with multiple metals in the host rock cannot be extracted economically at the mine site. The project cannot secure finance.*
- *Banks not willing to finance an innovative new mineral processing technology that has not been yet tested on a commercial scale.*

Action plan:

- Work on projects with proven mineral processing technologies that commercial banks are willing to finance.
- Match the funding type to the process technology. Banks may be too risk averse for certain technologies, but some private equity funds are willing to fund for new disruptive mineral processing technologies.
- Explore modular mineral processing options for expansion.
- If no viable mineral processing can be undertaken, mothball project and retest the assumptions every year with new technology.

4.9.7 Lack of Water for Mineral Processing

Most mineral processing circuits require water. In the absence of water, many mineral processes cannot produce economic yields. In many jurisdictions, a water licence is required to commence with mining. Mining projects compete with communities for water, and a water management system is vital. Many investors require the mine to demonstrate access to water or a viable dry mineral processing method prior to commitment of funds.

Example: A mining project cannot close on funding as a result of not securing water supply for the mineral processing plant.

Action plan:

- Ensure that enough water for mineral processing has been secured prior to committing CAPEX to the project.
- Explore dry mineral processing technologies that can be deployed.

4.9.8 Downturn in Commodity Market

The market for the mineral product has changed, and prices have dropped. The project becomes uneconomic, and no mining capital is secured.

Example: The feasibility study of a large iron ore project is completed, the mining licence is secured. The term sheet with financiers has been negotiated. Between 2011 and 2014 iron ore prices drop significantly and the conditions precedent of the financing term sheet are not met, the investor aborts the project.

Action plan:

- For large-scale mining projects, forward sales of mineral production for a period of 3 years to allow for a stable pricing during the beginning capital intensive years.
- Avoid developing mines on spot prices only.
- Mothball project until commodity market turns positive.
- Mine developer to carefully position the production entry of their project, aiming to bring the project into production at the upswing of the commodity price.

4.9.9 No Minerals Sales Agreement Concluded

The mining project cannot secure minerals sales agreement for an offtaker at economic rates. Reasons can be various, ranging from commodity market dynamics, technical issues on the project, logistical costs, licensing matters or skills related.

Example: An anthracite mining project in a remote location fails to secure an offtake agreement for its anthracite product at economic rates for the project. The logistics path proved to be too costly. The mining project is mothballed.

Action plan:

- Always develop a mining project with the market in mind first.
- The orebody and the market need to be linked economically.

4.9.10 Logistics Path Uneconomic

Despite a viable geology and reserves, economic mining and processing and viable commodity prices, the project is unfinanceable as the logistics path for the mineral product is too expensive with the mine being too far from the market. This is particularly relevant for bulk commodity projects, which are reliant on an economic logistics path.

Example: A bulk commodity mine is established in a remote location. CAPEX is spent on establishing the mine infrastructure and processing plant, and then opencast mining commences. One year later, the mine is mothballed, and investors have lost their investment. The reason: The logistics path was underestimated in terms of costs and has rendered the project uneconomic.

Action plan:

- Test the logistics path assumptions at the outset of the project, with competitive quotations from logistics providers.
- If the project logistics are uneconomic, abort the project.

4.9.11 Unrealistic Shareholder Expectations

Some shareholders and mine developers have unrealistic expectations on the value of their project. The NPV of their mining project may be high in value, but if no investor is interested to invest, they need to simply adjust their expectations if the project requires financing.

Example: The shareholders of a mining project have unrealistic expectations on the valuation of the project due to large expenditures incurred in the project. No investor can meet the shareholder expectations, and no deal is concluded.

Action plan:

- Conduct an independent valuation of the investment proposal, considering recent comparable market transactions.
- Ensure that shareholder expectations in line with the independent valuation.
- Shareholders to be aligned prior to negotiation.

4.9.12 Skills: Mining Team Not Experienced

The best project will not be financed if the mining team does not have members that are experienced in the operation and development of mines.

Example: Entrepreneurs have secured a large-scale copper concession but cannot raise any capital for the development of the project, as there is no team member with skills in the development and operation of copper projects.

Action plan:

- Do not develop mining projects without experienced mining skills.
- Attract experienced mining professionals to provide skills for the development and operation of mining projects on a retainer and profit share basis. This will provide comfort to investors.

4.9.13 Mining Project Has Poor Community Relationships

A mining project that has poor community relationships risks industrial action and strife, with associated production stoppages and other material losses. Investors will not invest in mining projects with poor community relationships.

Example: A platinum mine fails to meet its employment promises to the community on which the mine is located. The resultant strikes and community unrest lead to production stoppages and financial losses, and no investors are found for expansion capital.

Action plan:

- Identify community leaders.
- Ongoing and frequent communication with community leaders to foster beneficial relationships between the mining company and the community. Identify shortcomings and areas of dispute.
- Key issues are inability to meet community expectations and local employment opportunities. These are all addressed in the social and labour plan.

4.9.14 Poor Environmental Management

Mining projects that have inadequate provision for rehabilitation or are in a setting with historically high environmental liabilities will struggle to find investors, who fear both the commercial and reputational risks associated with such projects.

Example: A gold mine has been mismanaged, it is mothballed and there are insufficient funds in the rehabilitation bond. The mine must be recapitalized, but no investor is secured as a result of the high environmental liabilities.

Action plan:

- Make environmental sustainability a key pillar of the mining business case.
- Determine the extent of rehabilitation liabilities and profit to be generated from the remaining reserves. If there is a mismatch, and profits cannot deal with rehabilitation liabilities, do not proceed with the development of the project.
- Negotiate with the host government to pass on the historic environmental liabilities to the state, if possible, so that the mining project is unencumbered.

4.9.15 Balance Sheet: Over-geared Mining Operation and Mining Contractors

Mining operations and mining contractors are often over geared, i.e. have a too high debt to equity ratio. They have incurred too high debt for mining equipment and processing plants and cannot fund any further expansion. Their balance sheet does not attract any investor.

Example: A mining contractor, operating multiple mining contracts, secures a further mining contract but cannot finance the mining fleet to service this contract by conventional financing means due to the high-gearing ratio.

Action plan:

- Explore innovative off-balance sheet financing solutions that do not affect the balance sheet, e.g. leasing.
- Explore subcontracting partnerships with equipment suppliers and contractors.

4.9.16 Funding Request Too High and Mismatched to Reserves

Mining investment involves big financial numbers. The investment size, however, needs to be matched to the size of reserves. If the reserves are not large enough to repay the capital investment, as per definition of reserve tail ratio, with enough buffer in place, mining investors will not invest.

Example: An industrial minerals operation intends to refurbish the mineral processioning plant. The CAPEX is too high and mismatched to the remaining reserves, and no finance is secured.

Action plan:

- Ensure that the investment size matches the reserve size.

4.9.17 Equity Capital Required

The prospective funders require the developers to put in more capital into the project, before they inject own capital. The developer is asked to put more skin in the game. Usually a futile ask because this is precisely why the developer is reaching for the funder.

Example: A mining project seeking project finance from a commercial bank. The mine developer has paid for the CPR and the feasibility study, all licences have been obtained, there is an offtake agreement and mining, and processing contractors have provided quotations. The bank is interested to provide project finance but as a condition precedent requires from the mine developer a 30% equity contribution.

Action plan:

- Define the equity contribution. As CAPEX is required for equipment and services, partnerships with mining and processing contractors can be considered for equity contribution purposes, where the contractor's equipment is 'equity contribution'.
- Source a strong equity partner to cover the 30% equity contribution, while the project is being developed. The equity partner could be an equity fund or a larger mining company.

4.9.18 Not Meeting Financial Hurdle Rate

The financial hurdle rate is the minimum IRR that is required to attract deployment of mining capital; it is usually +25%. Mining projects need to be ideally positioned in the lowed OPEX cost quartile and have attractive IRR's even when commodity prices are lower. Mining projects and operations that have positive IRR's during peak commodity prices might not meet the financial hurdle rates set by investors when commodity prices are low.

Example: An iron ore mining project with a completed feasibility study, fully permitted, has an attractive IRR of +25% at peak iron prices. During the negotiation offinancing term sheet, the iron ore price falls and the IRR drops to 15%, not meeting the financial hurdle rate. The financing process is aborted, and the project is mothballed.

Action plan:

- Focus on the development of mining project in the lowest cost quartile that can withstand changes to commodity prices.
- Conduct risk simulations on the economic model, changing the production, commodity prices, CAPEX and OPEX parameters by +10% and -10% and determine the robustness of the IRR. When does the IRR not meet the hurdle rate and how likely is the simulated scenario likely to occur?

4.10 Summary and Mining Capital Game Plan

4.10.1 Summary

This book provides methods and best practices for mine developers to raise large sums of capital to develop their mining projects or mining operations. It covers current mining capital industry dynamics, the process and activities of developing a mining business case that can be funded, the documents of marketing the mining business case and the financial, legal and psychological foundation to raise mining capital (Fig. 4.31).

Mining capital is required for exploration projects, feasibility studies, to develop mining projects and bring them into production, purchase mine equipment and recapitalize existing mining operations. This guide is intended for:

- Mine developers seeking finance for their mining project
- Investors intending to invest in mining projects and operations
- Mining and processing contactors requiring equipment and plants to execute contracts
- Host governments seeking to develop their mining sector
- Consultants and brokers of mining finance

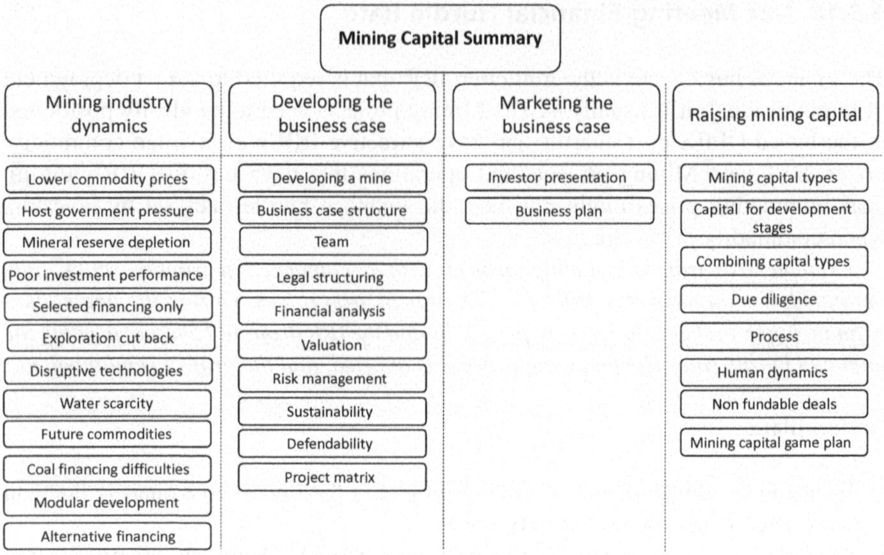

Fig. 4.31 Mining capital summary (Source: The author)

Mining Capital Industry Dynamics

In the current business environment, it is challenging to raise capital for mining projects and operations. In recent times, the mining capital sector is characterized by the following:

- Lower commodity prices with slow recovery
- Host governments applying pressure to mining companies
- Mineral reserve depletion and quality mining projects are harder to find
- Disruptive impact of the digital age on mining processes
- Water scarcity a factor for the mining business
- Focus on minerals for emerging technologies
- Poor historical performing of mining investments

This is, in turn, has affected the ability to finance mining projects conventionally. There is a subdued interest to invest in mining projects, only selected projects get funded, listings have been cancelled, and the share prices of listed juniors have followed a trend of growth and decline, commercial banks have reduced their resources teams, mining companies are overleveraged and the coal sector finds it difficult to raise capital.

At the same time, it has become easier to finance smaller mining projects than large mining projects, capital is redeployed into existing operations, producers have easier access to capital, and capital has become commodity specific.

In this environment, miners have resorted to alternative financing, with commodity traders, vendor financing, development financing, a modular implementation approach, collaboration strategic partnerships, and sale of assets as a source of finance. An important consideration for future mining financing is millennials and their approach to mining investment.

The solution lies in having a game plan to secure mining capital. This is the purpose of this guide, alternative and creative ways to finance mining projects, because of the difficulties of securing traditional sources of mining capital.

Developing a Mining Business Case for Investment: Methods

In the current business environment, only a strong and solid mining business case will raise capital. In Chap. 2, we covered the components, activities and required background information of developing a mining business case that will attract capital:

- Development of a mine—from exploration to production
- The mining business case structure—from orebody to market
- Team—a strong team will convince investors
- Legal structuring—the mining business case needs a solid and simple legal structure
- Financial analysis of the mining business case—the development of an economic model
- Valuation of mining projects—to enable negotiation on terms and conditions of the investment
- Risk management—mining is risky business; risks need to be identified and managed
- Environmental and social sustainability—with a poor legacy, more than ever do mining projects need to be environmentally and socially compliant to attract capital.
- Implementing the mining business case—plan versus reality
- Defendability—can the mining business hold up to investor scruitiny, is it defendable?

With a strong mining business case, the mine developer is ready to approach the market for capital.

The mining business case will be subject to an intense due diligence by investors. Resources are quantified; reserves are mined and processed to produce mineral products. The mineral products are transported to the customers via a logistics path. The mining business case must cover:

- Licences and host country dynamics
- Legal structure
- The team
- Exploration
- Mine plan
- Process design
- Infrastructure design

- Contractors and operators
- Offtake agreement
- Logistics path
- Environmental plan
- Social plan
- Product offtake agreement
- Project plan
- Economic analysis with key indicators NPV and IRR

The mining business case needs to be profitable and sustainable to attract mining capital. Many mining projects struggle to reach full funding capacity due to a shortcoming of the mining business case. The project matrix is the appropriate way to summarize the business case and identify the status, shortcomings and opportunities of the mining business case.

Marketing the Mining Business Case: Best Practice
The business case is presented in two key documents, the business plan and the investor presentation, outlined in Chap. 3. Once these have been understood, further due diligence on the feasibility study and related technical and commercial data is undertaken. The business plan and the investor presentation need to be simple, concise and paint the investment and return picture to investors. They are the key documents that are required to persuade the investors to invest their own capital into the mining business case.

Two examples are provided, which can be used as a reference guide and can be amended to suit the developers individual mining investment opportunity.

- A mining investment presentation of a gold and silver mining project in South America, fully licensed and feasibility study completed, seeking CAPEX to develop the project.
- The business plan of a vanadium—titanium—iron ore greenfields project that is seeking to raise CAPEX to implement the project in a phased manner, from pilot plant operations and mine design, to the implementation of mining, processing and toll smelting.

Raising Mining Capital: Best Practices
This chapter covered types of mining capital, capital for the different mine development stages, combining mining capital types, the process of due diligence, term sheets, the process of raising capital, human dynamics of mining capital, non-fundable mining deals and action plans and a mining capital game plan.

Mining Capital Types
There are different types of mining capital that are applicable to the different mine life stages, from exploration, feasibility, development to operations. The types of capital are:

- Equity finance
- Farm-in finance
- Listing
- Reverse listing
- Project finance
- Development finance
- Corporate bond finance
- Streaming finance
- Royalty finance
- Offtake finance
- Insurance wrapped finance
- Contractor finance
- Equipment finance
- Disinvestment finance

Due Diligence

The due diligence study is a key event on the road to securing mining capital. The prospective investor needs to undertake a thorough due diligence study on the mining investment proposal. The due diligence covers all key aspects that are covered in the business case with all supporting documentation and including the feasibility study. A successful due diligence can lead to adequate funding.

Term Sheets

A term sheet sets out the terms under which the mining investment will be made and serves as a template to develop the detailed legal agreement. Once the parties involved reach an agreement on the details laid out in the term sheet, a binding agreement is drawn up, prior to release of the funds into the project. Term sheets for the different types of mining investment have been presented, which can be used as a reference for transactions for the different types of mining capital.

The Process of Raising Capital

The process of raising mining capital follows ten steps. It starts with the mining business case and ends with the electronic deposit into the miner's account, enabling the project to be developed, at which point in time the mission is accomplished. The ten steps are:

1. Mining business case
2. Funding scope
3. Select and approach capital source
4. Investor presentation
5. Business plan and data
6. Due diligence
7. Investment decision
8. Negotiate term sheet
9. Agreement
10. Investment

Human Dynamics of Mining Deals

Raising capital for projects is also about people and psychology. Developers need to be aware of men of straw, people who give an impression of money, but where no money flows. Asking for proof of funds is a must. Investors are always attracted to a neat, simple and colourful packaged deal. There is always a difference in the perception of value, and common ground needs to be found to conclude a transaction. If, despite all intent, no common ground is found, deal fatigue sets in and the project may have to be shelved. The importance of a network cannot be overemphasized as a key component for success. 'Skin in the game' is a term often used by investor, tactically, referring to the equity contribution of the developer. Most important is a network. Mining deals are often concluded by members within a network. Mine developers need to be able to handle rejection. As many prospective investors will decline an opportunity, mine developers need to 'keep on walking'. A workable deal is a deal where both parties—the developer and the mining investor—have had to compromise and find the middle path to conclude the transaction. Mine developers need to be able to detect a 'poker face' as many investors will purposefully neither show positive nor negative emotions about a prospective investment opportunity. Perception management is important, and the funder needs to perceive that the development team can deliver on all aspects listed above and have a management plan in place to deal with the uncertainties prior to releasing funds. The crux for the developer lies in communicating the mining investment opportunity in simple terms to people who are new to mining. Finally, 'herd mentality' is very applicable to mining investments. If one investor decides to invest in a mining project, other investors follow suit, as they assume the due diligence has resulted in a positive outcome with good investment returns.

Non-fundable Mining Deals and Action Plans

The mining business cases cannot be financed when it is flawed, and mine developers and promoters need to recognize that. During the boom time of commodity prices, it was easier to finance mining projects with less checks and balances in place. These days this is not the case. The main reasons for poor business cases are outlined and action plans to address these are outlined. These include:

- Downturn in commodity market:
- No minerals sales agreement concluded
- Logistics path uneconomic
- Unrealistic shareholder expectations
- Mining team not experienced
- Mining project with poor community relationships
- Mining projects with poor environmental management
- Over-geared mining operations and mining contractors
- Funding request too high and mismatched to reserves
- Equity capital required
- Not meeting financial hurdle rate

4.10.2 Mining Capital Game Plan

The Mining Capital Game Plan (Fig. 4.32) outlines visually all key aspects of this guide. Four key elements form part of this game plan:

1. The mining business case
2. The mining capital types, with term sheets and transaction structures
3. The process of raising mining capital
4. The human dynamics

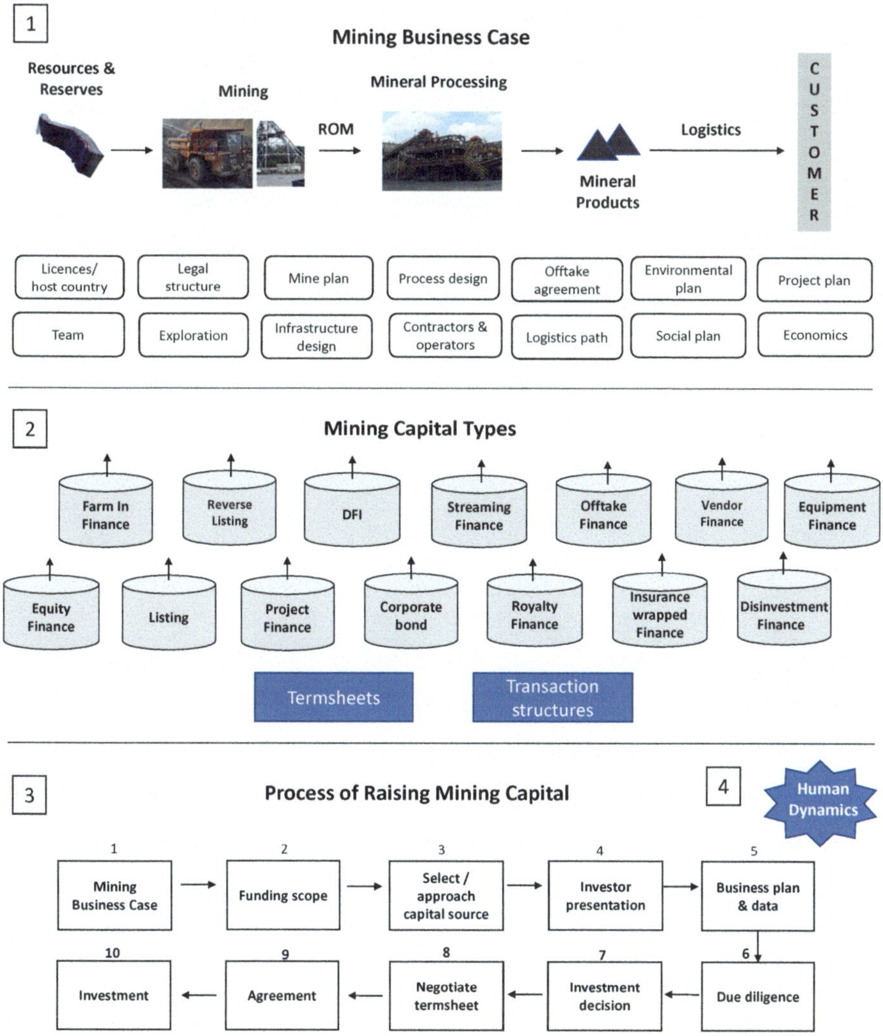

Fig. 4.32 Mining capital game plan (Source: The author)

I propose that mine developers use this mining capital game plan as a reference, together with this guide during their mining capital raising campaign.

A final note: Please ensure that all new and existing mines are environmentally and socially sustainable. This is a key criterion for mining investors. And future generations will hold us to account!

Good luck to all mine developers and mine owners in your capital raising initiatives. I hope, this guide will have provided you with new insights and information that you can use in your endeavours.

Mining Capital Case Studies

5

Contents

M. Seeger, *Mining Capital*, https://doi.org/10.1007/978-3-030-31225-1_5

5.1 Introduction to Mining Capital Case Studies

In this chapter, mining capital case studies are presented to illustrate details of mining capital transactions. We unpack the mining business case and the applicable funding plan, with the applicable mining capital type, term sheet and the transaction structure (Fig. 5.1).

The mining business cases outlined vary from exploration, feasibility, development, construction or operations.

Most case studies outlined are real and have been executed or have been tabled for consideration by the funders (Table 5.1). Some case studies have been modified and simplified for illustration purposes. Due to confidentiality matters, the company name, project name, exact location or details of the project are not provided. Some case studies are hypothetical, close to the real deal and are presented for demonstration purposes. The case studies presented serve as a benchmark for the reader's own mining projects seeking finance.

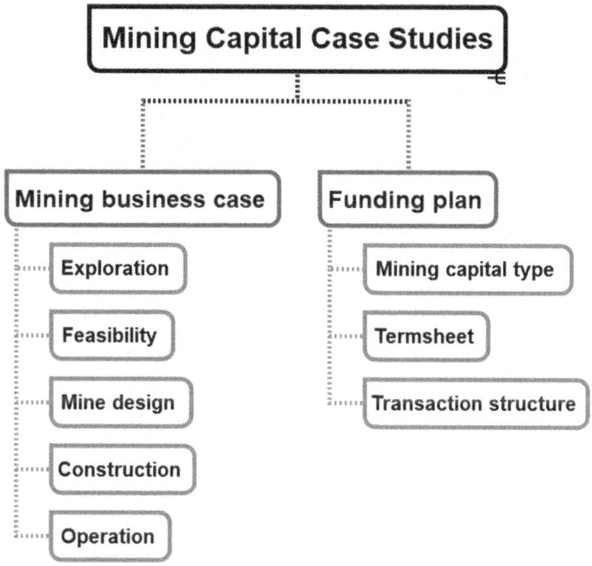

Fig. 5.1 Mining capital case studies—structure (Source: The author)

Table 5.1 Mining capital case studies

Mining business case	Funding requirement	Mining capital type
Copper project	$15 million	Equity finance
Gold project	$2 million	Farm-in finance
Phosphate project	$2 million	Listing
Diamond project	$2 million	Reverse listing
Coal mining project	$22 million	Project finance
Gold project	$10 million	Development finance
Titanium project	$40 million	Corporate bond finance
Gold project	$75 million	Streaming finance
Gold–silver project	$10 million	Royalty finance
Coal mine	$10 million	Offtake finance
Anthracite mine	$11 million	Insurance wrapped finance
Dolomite mine	$7 million	Contractor vendor finance
Mining contractor	$0.6 million	Equipment finance
Iron ore operation	$6 million	Disinvestment finance

Source: Authors' table

5.2 Equity Finance—Copper Project—$15 Million

5.2.1 Mining Business Case: Copper Project

The project is a copper project in South America that is to be implemented in three stages, seeking $15 million to commence with stage 1 of the project: mine design, site establishment, and construction. In stage 1, the project intends to 1,20,000 tpa of 25% copper concentrate (Table 5.2).

Table 5.2 Mining business case—copper project—$15 million

#	Item	Description
1	Project stage	• Mine development • Three-staged development approach
2	Location	• South America • 35 km from regional mining town
3	History	• Extensive copper production in the region over past 40 years
4	Corporate structure	• Holdco, a listed company, owns 85% of SPV • State-owned company owns 15% of SPV • SPV owns 100% mining licence
5	Development team	• Junior mining company
6	Permits and licences	• Mining licence
7	Geology	• Porphyry copper deposit
8	Exploration	• Exploration drilling programme completed
9	Resources and reserves	• Resources: 50 Mt., 4% copper, 0.2% cobalt • Mineral reserves: 3 Mt., 4% copper, 0.2% cobalt
10	Mining	• Mining method: opencast, truck and shovel • Contractor mining operations • Mining of the oxidised portion of orebody: 15 Mt. @ 2% Cu
11	Processing	• Process plant: Crushing, scrubbing, heavy media separation, solvent extraction electrowinning • Production capacity: 1,20,000 tpa • Product stage 1: copper concentrate, 25% Cu • Recovery rate of 85%
12	Production parameters	• Life of mine: 10 years • Stage 1: 1,20,000 tpa of 25% copper concentrate • Stage 2: 50,000 tpa copper cathode • Stage 3: 1,00,000 tpa copper cathode
13	Infrastructure	• Infrastructure to be developed: – Mine access road – Fence – Mine offices – Workshop – Stores – Water supply – Pollution control dam – Power supply – Communications and IT systems – Power supply – Fuel storage facility – Explosive magazine
14	Products	• Product: copper concentrate, 25% Cu • Copper price: $3.00/lb
15	Markets	• Product sold through an offtake agreement with a copper trading company
16	Logistics	• Road transport of copper oxide concentrate and copper cathode to smelter
17	Key suppliers	• Mining contractor • Processing contractor • Logistics contractor

(continued)

Table 5.2 (continued)

#	Item	Description	
18	Human resources	• Total work force: 200	
19	Environmental and social	• Environmental impact assessment approved	
20	Financial	Key financial indicators	
		CAPEX	$160 million (stages 1, 2 and 3)
		OPEX	Mining: $0.3/lb Processing: $0.5/lb Support services: $0.2 /lb Logistics: $0.4/lb
		NPV	+$300 million
		IRR	+35%
21	Project plan	• Total implementation period: 55 months – Stage 1: 15 months – Stage 2: 20 months – Stage 3: 20 months	
22	Funding requirements	• $15 million: – Complete mine design – Commence with construction of stage 1 of the project	

Source: Authors' table

5.2.2 Mining Capital Term Sheet and Transaction Structure—Copper Project

Term Sheet

The mine developer opted for *Equity Finance* and approached a mining fund to raise $15 million in exchange for shares in the SPV and structured the term sheet accordingly (Table 5.3).

Table 5.3 Term sheet equity finance—copper project—$15 million

Investor	• Mining fund
Investment	• $15 million
Shares	• 150 million new fully paid ordinary shares • This equates to 15% shareholding in the SPV
Share price	• $0.10/share
Terms	• The investment is for financing stage 1 of the project: Mine design, site establishment, construction • The mining fund will have the right to nominate one person as a director of company • Warrantees are provided by the project SPV
Conditions precedent	• No breach of warranties • Performance by SPV • Warranties given by the SPV remain true and correct • Compliance by the SPV with the social, environmental, health and safety policies that have been approved by • Mining fund will not be the single largest holder of shares in the company
Termination rights	• The mining fund may terminate by way of notice in the event of: – The Holdco or SPV being insolvent – A material breach of the warranties – If the number of ordinary shares allocated to mining fund would be less than 5% of the shares in the SPV

Source: Authors' table

Transaction Structure

The transaction structure depicts the mining fund investing $15 million equity in exchange for 13% in the SPV holding the mining licence. The SPV provides warrantees to the mining fund (Fig. 5.2).

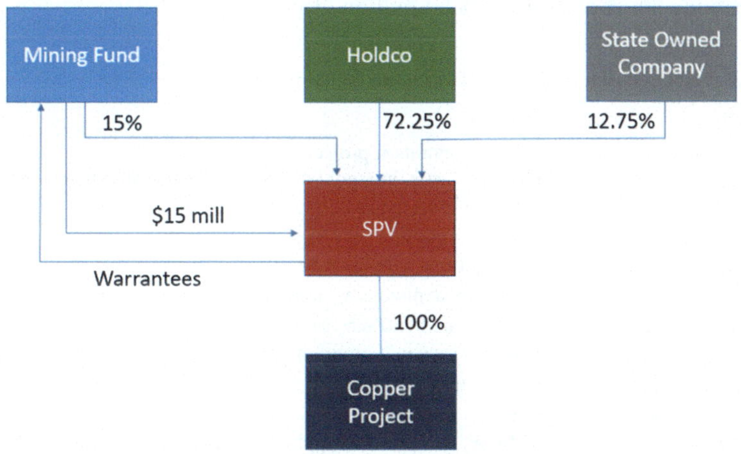

Fig. 5.2 Transaction structure—equity finance—copper project (Source: The author)

5.3 Farm-in Finance—Gold Project—$2.2 Million

5.3.1 Mining Business Case—Gold Project—$2.2 Million

The project entails a gold exploration project in Asia seeking to finance the concept and feasibility study at a cost of $2.2 million (Table 5.4).

Table 5.4 Mining business Case—gold project—$2.2 million

<table>
<tr><th>#</th><th>Item</th><th colspan="6">Description</th></tr>
<tr><td>1</td><td>Project stage</td><td colspan="6">• Gold exploration project
• The project is in a gold district, and numerous shallow artisanal works exist in the area.</td></tr>
<tr><td>2</td><td>Location</td><td colspan="6">• Asian country
• 30 km from a regional town</td></tr>
<tr><td>3</td><td>History</td><td colspan="6">• Multiple exploration campaigns conducted over 3 years</td></tr>
<tr><td>4</td><td>Corporate structure</td><td colspan="6">• SPV owned 100% by Holdco
• Holdco owned by four entrepreneurs</td></tr>
<tr><td>5</td><td>Development team</td><td colspan="6">• Junior mining company</td></tr>
<tr><td>6</td><td>Permits and licences</td><td colspan="6">• Prospecting licence
• Mining licence to be applied for once the feasibility study is complete.</td></tr>
<tr><td>7</td><td>Geology</td><td colspan="6">• Mineralisation is hosted within clays, metavolcanics and fresh mafics.
• The resource has 700 m strike extent
• The mineralisation dips 50° and varies in thickness between 30 and 10 m.
• Mineralisation defined by 0.5 g/t cut-off
• Extensive weathering</td></tr>
<tr><td>8</td><td>Exploration</td><td colspan="6">• 45 RC surface holes drilled
• Deposit tested with RC drill holes inclined at 55°
• Deposit estimated in a mine block model using the inverse distance square interpolation.
• The resource meets the criteria of the JORC code</td></tr>
<tr><td rowspan="4">9</td><td rowspan="4">Resources and reserves</td><td colspan="6">• JORC code</td></tr>
<tr><td colspan="2">Inferred</td><td colspan="2">Indicated</td><td colspan="2">Total</td></tr>
<tr><td>Tons</td><td>Gold g/t</td><td>Tons</td><td>Gold g/t</td><td>Tons</td><td>Gold g/t</td></tr>
<tr><td>6,00,000</td><td>2</td><td>8,00,000</td><td>3</td><td>14,00,0000</td><td>2.5</td></tr>
<tr><td>10</td><td>Mining</td><td colspan="6">• Mining method: Opencast mining
• Production rate: 1,20,000 tpa
• Stripping ratio: 3:1
• Au grade: 2.5 g/t
• Life of mine: +7 years</td></tr>
<tr><td>11</td><td>Mine equipment</td><td colspan="6">• All equipment supplied by mining contractor</td></tr>
<tr><td>12</td><td>Processing</td><td colspan="6">• Planned CIL gold plant includes:
– Ore receiving and crushing
– Milling and classification
– Gravity concentration</td></tr>
</table>

(continued)

Table 5.4 (continued)

#	Item	Description
		– Pre-leach thickener – Leaching and adsorption – Cyanide detox and final tailings disposal – Elution – Electrowinning – Gold room – Carbon regeneration and acid wash
13	Production parameters	• Not established
14	Infrastructure	• Mine infrastructure to be established: – Process plant – Mine access road – Security – Offices – Change house – Stores – Workshop – Tailings dam – Waste disposal area – Pollution control dam
15	Products	• Gold doré
16	Markets	• Gold sold to gold international refinery
17	Logistics	• Transport of gold via helicopter and silver • Security company appointed to transport gold bars to customer
18	Human resources	• The total workforce of the planned mine is 150.
19	Environmental and social	• EIA to be applied for
20	Financial	• Concept study: $0.2 million • Feasibility study and drilling: $2 million • CAPEX: $12 million • OPEX: $16/t • Indicators: NPV: $10 million, IRR 30%, payback period 3 years.
21	Project plan	• Complete JORC report • Project permitting • 5000 m drill programme • In-house resource estimation • Concept study • Feasibility study
22	Funding requirements	• Phase 1: $ 0.2 million to complete the concept study • Phase 2: $2.0 million to complete the feasibility study • Phase 3: The project CAPEX, $10 million, will be acquired via a bank loan once the feasibility study is completed.

Source: Authors' table

5.3.2 Mining Capital Term Sheet and Transaction Structure—Gold Project—$2.2 Million

Term Sheet

The mine developer secured finance from an investor via a *Farm-in Finance* deal. The strategic investor, a mid-tier gold mining company, acquired in two phases: 50% of the SPV holding the project by providing financing for the concept and feasibility study, and further secured an option to increase shareholding the project (Table 5.5).

Table 5.5 Term sheet farm-in finance—gold project—$2.2 million

Developer	• Holdco: Owns 100% of SPV • SPV: Owns the prospecting permit and requires equity funding to develop the project
Investor	• A gold mining company that wishes to acquire shares in the SPV • Willing to invest into the SPV in staged phases, through a farm-in into the SPV, acquiring shares in phases
Investment requirements	• Phase 1: $0.2 million to complete the concept study • Phase 2: $2 million to complete the feasibility study • Phase 3: $10 million to construct the mine
Conditions precedent	• SPV allows the investor to conduct a legal and technical due diligence • The investor has the exclusive right to acquire shares in the SPV
Payments	• The investor pays the SPV $0.2 million as part of acquiring 15% of the prospecting permit which will be used to complete the concept study • Investor pays the SPV $2 million as part of acquiring 50% of the prospecting permit to pay for the feasibility study
Shares	• Phase 1: Upon the conclusion of the concept study, Holdco issues investors 15% of the ordinary shares of SPV • Phase 2: Upon the completion of the feasibility study, Holdco issues investor a further 35% of the ordinary shares of SPV. The investor will have acquired 50% of the issued shares by investing $2.2 million in the SPV
Timelines	• Payment 1 must be executed within 10 days of signing the farm-in agreement • Payment 2 must be executed within 2 years of signing the farm-in agreement
Option	• Holdco grants investor the right to acquire the remainder of shares of the SPV at market-related prices, determined by independent valuator
Warrants	• Holdco warrants to the investor that the prospecting licence – is correctly registered – in good standing – has not been offered – has no environmental claims

Source: Authors' table

Transaction Structure

The transaction structure (Fig. 5.3) depicts the investor, the gold mining company, acquiring shares in the SPV holding the gold project in two phases. Phase 1 entails investing $0.2 million for the concept study in exchange for 15%. Phase 2 entails investing an additional $2 million for the feasibility study for an additional 35% in the project, totalling 50% ownership of the SPV.

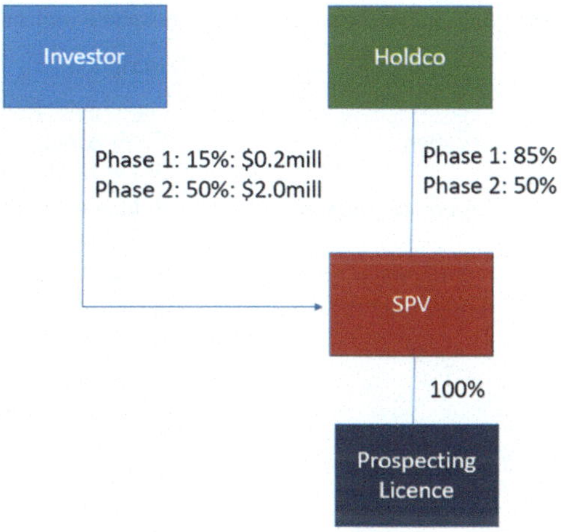

Fig. 5.3 Transaction structure—farm-in finance—gold project (Source: The author)

5.4 Listing—Phosphate Project—$2 Million

5.4.1 Mining Business Case—Phosphate Project—$2 Million

The project entails a phosphate exploration project seeking to finance the exploration campaign at a cost of $2 million (Table 5.6).

Table 5.6 Mining business case—phosphate project—$2 million

#	Item	Description
1	Project stage	• Exploration project
2	Location	• North Africa • 50 km from regional town
3	Corporate structure	• Holdco owns 100% of SPV • SPV owns 100% prospecting permit
4	Development team	• Junior mining company
5	Permits and licences	• Prospecting licence
6	Geology	• The intrusive carbonatites complex, steeply dipping, volcanic source • Carbonite diameter: 1500 m × 200 m high • The carbonatite is a composite ring complex.
7	Exploration	• 1970s: – Rock chip sampling campaign, 1000 samples on a 200 m × 10 m grid, defined trends – Five holes to a depth of 70 m each • Planned: – Reassessment of historical drill cores – Helicopter-borne geological survey – Rock sampling 11 m × 50 m based on helicopter data – Drilling of 10 holes to a depth of 200 m each – Development of 3—D geological model – Cut-off grade: 5% P_2O_5
8	Resources and reserves	• Inferred resources: 110 million tons • Grade: 5% P_2O_5 • Concentrate grade: 32% P_2O_5
9	Mining	• Mining method: Opencast, truck and shovel • Depth of mining: 150 m • Contractor mining operations: drill, blast, load and haul • Life of mine: +20 years
10	Processing	• Metallurgical testwork: • 1970s: Hydro metallurgy testwork undertaken on oxidised surface material • Planned: Testwork on fresh un-oxided material • Planned mineral processing circuit: crushing, milling, flotation, acid leaching, granulation • Target recovery grade: +80% • Target concentrate grade: +30% P_2O_5
11	Production parameters	• To be determined during the feasibility stage • Planned: 5 Mtpa ROM ore

(continued)

Table 5.6 (continued)

#	Item	Description	
12	Infrastructure	• Good infrastructure near phosphate project – Grid power – Tarred road in close proximity – Railways station 100 km from site – Port 400 km from site • Infrastructure to be developed: – Site roads mine complex infrastructure	
13	Products	• Phosphate concentrate	
14	Markets	• Sold through an offtake agreement with an international fertiliser company • P_2O_5 concentrate price: $125/ton	
15	Logistics	• Road transport of phosphate concentrate 100 km • Rail transport of phosphate concentrate 400 km • Shipment of product to customers	
16	Human resources	• To be determined during the feasibility stage, estimate 200	
17	Environmental and social	• Environmental impact assessment will be conducted as part of mining licence application	
18	Financial	• Key financial indicators, to be confirmed during feasibility study:	
		CAPEX	$150 million
		OPEX	$35/t
		IRR	+25%
19	Project plan	• Total implementation period: 3 years – Stage 1: Exploration, feasibility study and permitting: 2 years – Stage 2: Construction phase: 1 year	
20	Funding requirements	• $15 million to finance stage 1 of the project	

Source: Authors' table

5.4.2 Mining Capital Term Sheet and Transaction Structure—Phosphate Project—$2 Million

Term Sheet

The shareholders of the phosphate exploration project decided to take their company (Holdco) to market and *list it on a stock exchange* to raise the required capital for Stage 1 of the project. The terms of the listing are outlined in Table 5.7.

Table 5.7 Term sheet listing—phosphate project—$2 million

Company listed	• Phosphate Company Holdco PLC
Valuation	• Independent valuation determined the phosphate project value to be $5 million
Investment	• $2 million
Shares	• 5,00,00,000 new fully paid ordinary shares
Share price	• $0.10/share
Terms	• The investment must be used for financing the exploration, feasibility study and permitting of the phosphate project
Shares structure	• Issued shares: 5,00,00,000 shares • Holdco shareholders (developer): 3,00,00,000 shares • Public shareholders: 2,00,00,000 shares
Holdco shareholders	• Issued 3,00,00,000 shares at $0.01/share for vending their project into Phosphate Company PLC • This equates to 60% of the company
Public shareholders:	• Paid $0.10/share for 2,00,00,000 shares
Share price	• $0.10/share
Shareholding	• Holdco shareholders 60% • Market 40% – Institutional investors 30% – Retail investors 10%
Advisor	• Sponsoring broker X appointed as the nomad • Reputable advisory company appointed as corporate financial advisor
Independent mining consultant	• Independent mining consultant to conduct due diligence report on phosphate project and provide valuation • Total value of project: $5 million
Legal firm	• To conduct due diligence on company, secretarial documentation and agreements
Documentation	• Prospecting licence • Competent persons report on the phosphate asset • Holdco company secretarial documentation • Audited financials of Holdco • Holdco team member's CVs • Proof of expenses • Project plan • Information memorandum on phosphate investment, outlining phosphate business case
Escrow period	• Holdco shareholders may not sell their shares for 2 years
Directors	• Three executive directors nominated from Holdco • Two non-executive directors

(continued)

Table 5.7 (continued)

Timing	• 4 months
Fees	• Holdco shareholders to pay $2,00,000 for listing process
Annual stock exchange fees	• $2,00,000
Reporting	• Phosphate company PLC to publish: – Announcement after drilling campaigns – Quarterly reports – Annual reports
Shareholders meeting	• To be held annually

Source: Authors' table

Transaction Structure

The public shareholders, entailing institutional and retail investors, are issued 20 million shares, at $0.10/share and Holdco shareholders are issued 30 million shares as at $0.01/share, for work done to date on the project. Holdco shareholders will benefit significantly should the share price appreciate. Through this process, PLC raised $2 million which is used to advance the project (Fig. 5.4).

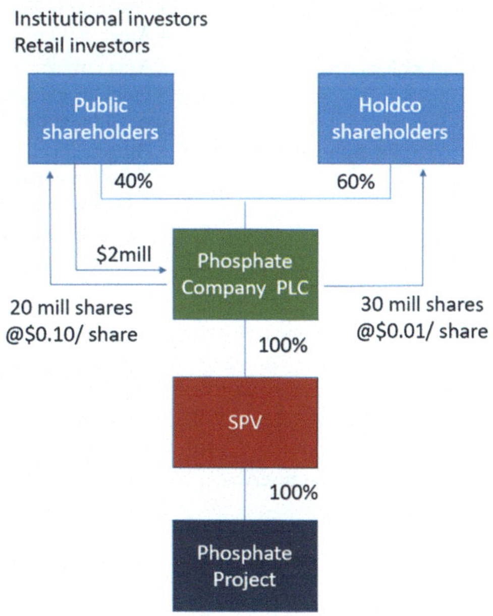

Fig. 5.4 Transaction structure—listing—phosphate exploration company (Source: The author)

5.5 Reverse Listing—Diamond Project—$2 Million

5.5.1 Mining Business Case—Diamond Project—$2 Million

The project entails a diamond exploration project seeking to finance the exploration campaign at a cost of $2 million (Table 5.8).

Table 5.8 Mining business case—diamond project—$2 million

#	Item	Description
1	Project stage	• Diamond project, feasibility stage • Feasibility study to be started, followed by trial mining and mine development
2	Location	• America • 150 km from regional town
3	History	• Several large mining companies conducted exploration on tenement area
4	Corporate structure	• Holdco owns 100% of SPV • SPV owns 100% prospecting licence
5	Development team	• No mine development team appointed
6	Permits and licences	• Prospecting licence • Area: 200 km^2
7	Geology	• Area covered by recent sediments and sands • Previous exploration campaign confirms presence of one large 8 ha kimberlite
8	Exploration	• Historical activities – Aeromagnetic surveys – Ground magnetic surveys – Soil geochemistry surveys – RC drilling and core drilling • Planned activities – 5000 m of core drilling – 5000 m of large diameter drilling – 2000 kg of micro-diamond analysis – 1000 tons of macro-diamond analysis – Geophysical surveys – Competent persons report
9	Resources and reserves	• Potential kimberlite resource: 35 Mt • Grade: 40 cpht at $30–300/carat
10	Mining	• Mining method: opencast, truck and shovel • Stage 1: Trial mining operation of 5,00,000 tons of kimberlite ore • Stage 2: Planned production rate: 1 Mtpa ROM • Potential life of mine: +20 years
11	Processing	• Planned: – Modular DMS plant: Scrubbers, DMS and screening plant, double tier X diffraction concentrators, grease table, diamond recovery plant/sort house, thickener – Production rate: 150 tph

(continued)

Table 5.8 (continued)

#	Item	Description
12	Production parameters	• To be finalised during the feasibility stage • Planned production: 1 Mtpa kimberlite ore
13	Infrastructure	• Limited infrastructure near site – No grid power – Tarred road 80 km from site – Full mine complex, with off grid power to be developed
14	Products	• High-quality coloured diamonds, $150/ct
15	Markets	• Offtake agreement with international diamond trader • Key driver: Jewellery demand
16	Logistics	• Diamonds to be air transported to diamond trader
17	Human resources	• To be determined during the feasibility stage, estimate 120
18	Environmental and Social	• Environmental impact assessment will be conducted as part of mining licence application
19	Financial	• Indicative financial indicators, to be confirmed during feasibility study:
		CAPEX: • Stage 1: Exploration and feasibility: $2 million • Stage 2: Trial mining: $10 million Stage 3: Large-scale operation: $120 million
		OPEX: $25/t
		IRR: +22%
20	Project plan	• Total implementation period: 3 years – Stage 1: Feasibility study: 1 year – Stage 2: Trial mining: 1 year – Stage 3: Large mine development: 1 year
21	Funding requirements	• $2 million equity finance to assist in funding stage 1 of the project (bulk sample campaign and feasibility study)

Source: Authors' table

5.5.2 Mining Capital Term Sheet and Transaction Structure—Diamond Project—$2 Million

Term Sheet

The shareholders of the diamond exploration project decided to opt for *Reverse Listing Finance* and vend their project into a listed company on a premier mining stock exchange to enable the financing of Stage 1 and 2 of the project (Table 5.9).

Table 5.9 Term sheet reverse listing—diamond project—$2 million

Project	• Diamond exploration project
Valuation	• Independent valuation determined the diamond exploration project value to be $2 mil
Shareholders	• Holdco, 100% owner of diamond company SPV
Project company	• Diamond company SPV
Investment required	• Stage 1: $2 million
Listed company listed	• Diamond Company PLC • Cash reserves: $3 million
Market cap before project vend in	• $10 million
Share structure before project vend in	• Issued shares: 100 million shares issued • Share price: $0.10/share
Shares issued to Holdco shareholders	• Valuation of Holdco: $2.5 million • Shares issued: 25 million shares issued • Share price: $0.10/share
Market cap of diamond company PLC post vend in	• $12 million
Commitment by PLC	• Use of $2 million from Diamond Company PLC cash reserves to develop stage 1 of project • Raise capital on stock exchange to develop stage 2 of project ($10 million)

Source: Authors' table

Transaction Structure

Holdco vends in its shares, valued at $2.5 million into listed Diamond Company PLC (Fig. 5.5).

Holdco shareholders are issued 25 million shares at 10c/share. Diamond Company PLC adds these to the 100 million issued shares, and the issued shares are 125 million.

As part of the deal, Diamond PLC committed to financing the feasibility study cost at $2 million.

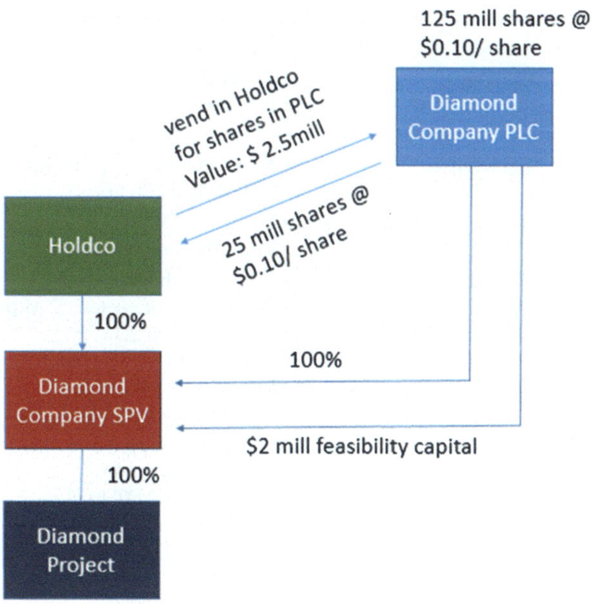

Fig. 5.5 Transaction structure—reverse listing—diamond project (Source: The author)

5.6 Project Finance—Coal Project—$22 Million

5.6.1 Mining Business Case—Coal Project—$22 Million

The project entails a financing an opencast coal project. Exploration and the feasibility study have been completed, the project has a mining licence and there is an offtake for the thermal coal (Table 5.10).

Table 5.10 Mining business case—Coal project—$22 million

#	Item	Description	
1	Project stage	• Mine design • Opencast coal mining project	
2	Location	• Africa • Distance to nearest town: 25 km	
3	History	• Coal project was drilled in the 1970s by major energy companies	
4	Corporate structure	• Holdco owns 70% of the SPV • Local partner owns 30% of the SPV • Project SPV owns 100% of the mining licence	
5	Development team	• Mid-tier mining company	
6	Licence	• Mining licence	
7	Geology	• Thermal coalfield • Bituminous coal • 2 seams: 1 seam (3 m) and 2 seam (2 m)	
8	Exploration	• Major mining company drilled 30 holes • SPV drilled 100 holes	
9	Resources and reserves	• Resources: 50 Mt • Mineable reserves: 20 Mt	
10	Mining	• Mining method: opencast, truck and shovel • Production parameters: 2 Mtpa ROM • Stripping ratio 2:1 • Geological loss 5%, mining loss 5% • Life of mine: 10 years • Mining contractor operation	
11	Mine equipment	Mining fleet, supplied by mining contractor: • 3 × 85 t excavators • 12 × 40-ton ADT trucks • 1 × drilling machines • 2 × dozers D10 • 1 × grader • 2 × front end loader • 1 × explosive truck • 1 × water tanker • 1 × diesel bowser • Utility equipment: 4 × 4s, service vehicle, crane, mobile lights	
12	Processing	• Processing plant: – Crushing – Screening – Dense medium separation • Process plant to be outsourced to a processing contractor	
13	Production parameters	**Production plan**	
		Item	
		Mineable reserve	20 Mt
		ROM/annum	2 Mtpa
		Stripping ratio	2:1
		Geological losses	5%
		Mining losses	5%
		Yield	70%

(continued)

Table 5.10 (continued)

#	Item	Description	
14	Infrastructure	• Mine infrastructure to be developed: – Mine access road – Upgrading of railway siding – Workshop – Pollution control dam – Electrical infrastructure: Substation 5 MVA – Slimes dam/co-disposal system – Mine workshop – Mine offices	
15	Products	• The coal mine will produce export grade coal	
		Product	**Export grade coal**
		Yield	60%
		CV (MJ/kg)	25
		Volatiles	24
		Ash	16
		Sulphur	0.8
16	Markets	• Coal sales agreement secured with thermal power plant • Coal sold FOT (free on truck)	
17	Logistics	• Coal product sold free on truck (FOT)	
18	Human resources	• The total workforce of the mine: 200 • All employees live off-site	
19	Environmental	• EIA approved	
20	Social and community	• Social and labour plan completed and approved	
21	Financial	Key financial indicators	
		CAPEX	$30 million
		OPEX	$20/ton
		NPV (10%)	$25 million
		IRR	25%
		Payback period	4 years
22	Project plan	• Exploration: Year 1 • Feasibility study: Year 2 • Development: Year 3 • Construction and commission: Year 3 • Production: Year 4	
23	Funding requirements	• $22 million to fund the development of the mine • $8 million equity capital available	

Source: Authors' table

5.6.2 Mining Capital Term Sheet and Transaction Structure—Coal Project—$22 Million

The shareholders of the coal project decided to apply for *Project Finance* with a commercial bank, as all criteria for project finance were achieved: licences, surface agreements, CPR, mine plan, offtake agreement, economic logistics path, environmental approval, strong mining team (Table 5.11).

Table 5.11 Project finance—coal project—$22 million

Lender	• Commercial Bank
Borrower	• SPV
Shareholders	• Holdco 70% and local partner 30%
Sponsor	• Holdco
Obligors	• SPV, local partner and the sponsor jointly
Facilities	• Senior debt facility and standby debt facility
Senior debt facility	• $20 million to fund the capital expenditure of the project
Standby debt facility	• $ 2 million to fund operating cost overruns or working capital shortfalls
Standby equity facility	• The sponsor shall be required to cede a standby equity account containing no less than $2 million to the lender to fund any capital and/or operating cost overruns or working capital shortfalls, which account shall be released upon completion.
Ranking	• Senior
final maturity date	• 7 years from financial close
Financial close	• The date on which the lender notifies the borrower that the conditions precedent to financial close have been fulfilled in form and substance satisfactory to the lender.
Arranging and underwriting fee	• 2% of the facility amount to be paid in instalments on credit approval, signature of the finance documents and financial close
Availability period	• 2 years from financial close
Commitment fees	• 1.5% per annum
Interest rate	The sum of: • The base rate; and • The applicable margin
Base rate	• 3-month LIBOR
Default interest	• The interest rate plus 2%
Interest payments	• Quarterly
Principal repayments	• Capital is to be repaid quarterly commencing 36 months from financial close
Voluntary prepayment penalties	• Voluntary prepayments allowed subject to a 2% penalty.
Mandatory prepayments	The lender will have the right to request prepayment of the facilities in certain instances: • A change in control of any of the obligors; • The sale of all or substantially all the assets of the obligors; • Receipt of material insurance proceeds by the obligors; or • Performance in terms of the finance documents becomes illegal for any party thereto.
Project accounts	• The bank accounts to be opened for purposes of the project: – Proceeds account – Debt service reserve account – Distributions account
Debt service reserve account	• The borrower shall establish a debt service reserve account with the lender. • The balance in this account shall be maintained at an amount equal to the next 6 months' payments of capital and interest due under the facilities. • This account shall be pledged and ceded as security and will be used for meeting any shortfall in debt service under the facilities.

(continued)

Table 5.11 (continued)

<table>
<tr><td>Payment waterfall</td><td>All payments shall be subject to a conventional payment waterfall structure, as follows, in priority:
1. Taxes and royalties;
2. Fees and costs due under the facilities;
3. Operating and capital costs projected in any approved development plan;
4. Interest service on the senior debt facility;
5. Capital service on the senior debt facility;
6. Interest service on the standby debt facility;
7. Capital service on the standby debt facility;
8. Cash sweep; and
9. Discretionary transfers to the distribution account</td></tr>
<tr><td>Security</td><td>• Full security over all the assets and mining licence of the borrower
• Completion guarantees from the sponsor
• Limited guarantees from the shareholders secured by a cession and pledge of their shares and claims against the borrower.</td></tr>
<tr><td>Completion</td><td>• Completion is defined as the passing of technical, operational and economic completion tests which shall be developed in conjunction between the borrower, the lender and the independent technical consultant, who will be appointed by the lender and will be required to verify the satisfactory results of these tests.</td></tr>
<tr><td>Independent technical consultant</td><td>To be appointed by the lender and whose identity is to be acceptable to the borrower for the performance of duties to an agreed scope of work including, but not limited to:
• A high-level review of technical information for the project during the lenders' due diligence;
• The monitoring of construction progress before completion; and
• Verifying the cost to reach completion and performing completion tests.</td></tr>
<tr><td>Financial covenants</td><td>The following financial covenants shall apply and shall be measured semi-annually on both a historic and forward-looking basis:
<table>
<tr><th>Covenant</th><th></th><th>Plan</th><th>Default level</th></tr>
<tr><td>Debt service cover ratio</td><td>No less than</td><td>1.6</td><td>1.4</td></tr>
<tr><td>Project life cover ratio</td><td>No less than</td><td>1.8</td><td>1.6</td></tr>
<tr><td>Reserve tail ratio</td><td>No less than</td><td>30%</td><td>30%</td></tr>
</table></td></tr>
<tr><td>Distribution conditions</td><td>The borrower shall not pay, make or declare any distribution to shareholders unless:
• Project completion is achieved
• First principal repayment made
• The financial covenants are above the distribution levels as at the most recent calculation date</td></tr>
<tr><td>Conditions precedent</td><td>• Satisfactory conclusion of due diligence
• The sponsor's equity commitment to spend on project CAPEX
• Satisfaction with all material contracts, including cession and step-in clauses as applicable
• All licences to be in place
• Satisfaction with the project in respect of the equator principles.</td></tr>
</table>

Source: Authors' table based on project finance term sheets for mining projects

Transaction Structure

In the transaction structure, the bank provides a $22 million debt facility to the SPV. Holdco and the Local Partner provide completion guarantees to the bank (Fig. 5.6).

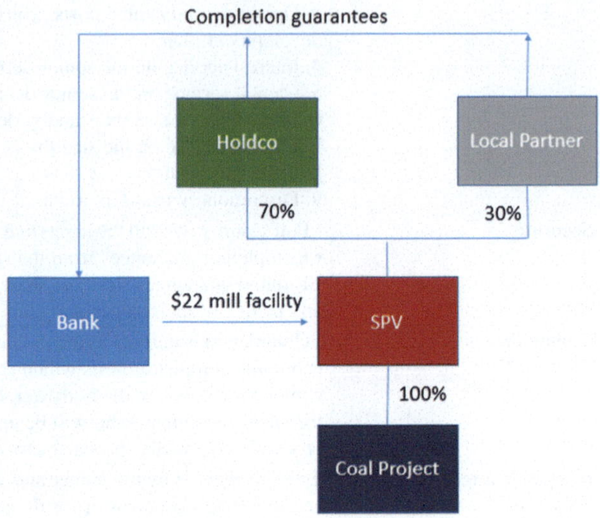

Fig. 5.6 Transaction structure—project finance—coal project (Source: The author)

5.7 Development Finance—Gold Project—$10 Million

5.7.1 Mining Business Case—Gold Project—$10 Million

The project entails a small-scale gold project, located in close proximity to large gold operations securing development finance to scale up operations (Table 5.12).

Table 5.12 Mining business case—gold project—$10 million

#	Item	Description
1	Project stage	• Operation • Small-scale gold and silver underground mine developed along a system of narrow veins • Owner intends to develop the operation into a 10,000 tpm operation producing gold and silver doré and concentrates. • Based on regional geology, potential for a large-scale mining operation
2	Location	• South America
3	History	• Mine operated by small-scale miners for several decades
4	Corporate structure	• Holdco owns 100% of the project SPV • SPV owns 100% of the licences
5	Development team	• Junior mining company

(continued)

Table 5.12 (continued)

<table>
<tr><th>#</th><th>Item</th><th>Description</th></tr>
<tr><td>6</td><td>Permits and licence</td><td>• Mining licence on 1700 ha
• Environmental permit
• Surface access agreement</td></tr>
<tr><td>7</td><td>Geology</td><td>• The mine is in a mineralised gold district that encloses multi-million-ounce gold deposits.
• Type of deposit: vein deposits
• Gold–silver mineralisation in four subparallel faults and brecciated zones, 100–150 m apart, filled by hydrothermal quartz, sulphides and oxides.</td></tr>
<tr><td>8</td><td>Exploration</td><td>• Exploration activities: geological mapping, surface and underground sampling, surveys, diamond drilling, metallurgical testwork, underground development, extraction of 1,10,000 ton bulk sample
• Drilling: nine holes drilled totalling 1600 m</td></tr>
<tr><td>9</td><td>Resources and reserves</td><td>Mineral resource statement:
<table>
<tr><th>Category</th><th>Tons (Mt)</th><th>Au grade (g/t)</th><th>Ag grade (g/t)</th></tr>
<tr><td>Measured and indicated</td><td>1</td><td>2.9</td><td>278</td></tr>
<tr><td>Inferred</td><td>0.5</td><td>3.3</td><td>198</td></tr>
</table></td></tr>
<tr><td>10</td><td>Mining</td><td>• Mining method: Shrinkage mining suitable for narrow structures
• Historic mining activities occurred along veins within mineralisation of the oxidised and transition zones.
• The mineral is transported to the mine portal, where it is discharged into hoppers for loading into 10-ton trucks for hauling to a stockpile at the plant site.</td></tr>
<tr><td>11</td><td>Mine equipment</td><td>• Mine equipment for underground gold mining method: Compressor, drilling machines, rail tracks, tools, hoppers, mining truck</td></tr>
<tr><td>12</td><td>Processing</td><td>• Mineral processing plant:
– Crushing, grinding and conventional cyanidation by heap leaching.</td></tr>
<tr><td>13</td><td>Production parameters</td><td>• Production rate: 60,000 tpa
– Gold recovery: 85%
– Silver recovery: 95%
• Life of mine: 14 years</td></tr>
<tr><td>14</td><td>Infrastructure</td><td>• Mine infrastructure required:
– Process plant
– Mine access road
– Security
– Offices
– Change house
– Stores
– Workshop
– Tailings dam
– Waste disposal area
– Pollution control dam</td></tr>
</table>

(continued)

Table 5.12 (continued)

#	Item	Description	
15	Products	• Doré—Hosts gold and silver • Slag: Crushed, kept in barrels for shipment to a smelter sale. • Concentrates: For shipment to a smelter	
16	Markets	• Customers: Buyers in the USA and China, smelters and refineries • Smelter charges: $250/dry ton • Pricing: Customer pays for 99% of the contained gold and silver.	
17	Human resources	• The total workforce of the mine: 150	
18	Environmental	• Environmental impact study approved.	
19	Social	• Social and labour plan includes village school refurbishment, payment of school personnel, health and development programme	
20	Financial	Key financial indicators	
		CAPEX	$10 million
		OPEX	$127/t
		NPV (10%)	$4.5 million
		IRR	20%
		Payback period	5 years
21	Project plan	• Mine design: 6 months • Construction and commissioning: 9 months • Operation: month 16	
22	Funding requirement	• $10 million to develop the small-scale mine	

Source: Authors' table

5.7.2 Mining Capital Term Sheet and Transaction Structure—Gold Project—$10 Million

Term Sheet

The gold project qualified for *Development Finance* from a regional development finance institution, as it was in an area with high unemployment. The community and secondary industry would benefit from this project. The mining project, although small, had the potential to unlock a much larger mining project, given its proximity to nearby multi-million-ounce gold deposits (Table 5.13).

Table 5.13 Term sheet development finance—gold project—$10 million

Lender	• Development finance institution
Borrower	• SPV
Guarantor	• Holdco
Amount of facility	• $10 million
Interest rate	• 9.25% pa
Tenor	• The tenor of the loans is 60 months
Loan documents	• The loan agreement includes: – Conditions to funding – Representations and warranties – Affirmative and negative covenants – Information covenants – Events of default – Indemnity and boilerplate provisions – Voting rights – Security sharing – Guarantees – Share retention – Subordination
Security	• Security will be first ranking security over: – All the company's assets – The SPVs bank account – Intercompany loans amongst SPV and the guarantors
Transaction documents	• Financing documents • Project documents • Letters of intent with respect to construction contracts and mining services agreement
Legal/compliance	• SPV and the guarantor to provide corporate authorisation documents
Miscellaneous	• Due diligence on the project • Project conditions: ore reserves report • Insurance: SPV must have adequate level of insurance in place and certification obtained from SPVs insurers • Financial model: An agreed financial model, which includes a mine plan • Environmental and social conditions: – Acompleted social and environmental assessment – An environmental and social action plan – Annual monitoring report – Implementation of an agreed S&E management system – Appointment of a third-party consultant

(continued)

Table 5.13 (continued)

Default event	• A default arises for SPV or the guarantors triggering one or more of the following key events of default: – Non-payment – Breach of representation or undertaking – Government expropriation or nationalisation – Insolvency – Guarantors' liabilities exceeding US$1 million – Authorisations not maintained or restored within 30 days of a notice requiring restoration – Non-performance of project documents – All or a material part of the secured property is destroyed, lost or damaged beyond repair

Source: Authors' table based on typical development finance term sheet

Transaction Structure

In the transaction structure the Development Finance Institution (DFI) provides a loan of $10 million to the SPV. Holdco is the guarantor to the DFI, and the assets of the gold project provide security to the DFI (Fig. 5.7).

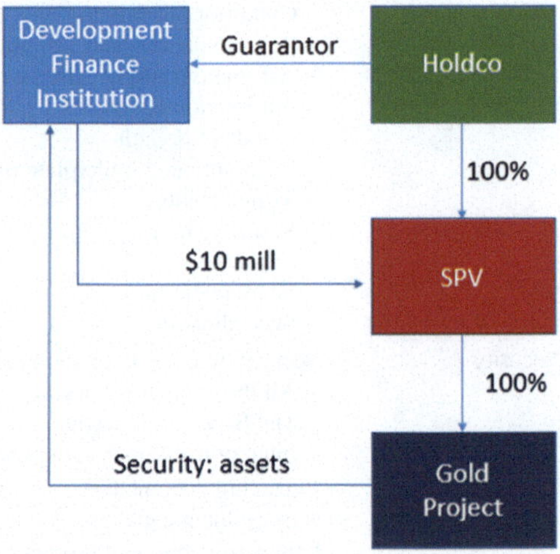

Fig. 5.7 Transaction structure—development finance—gold project (Source: The author)

5.8 Corporate Bond Finance—Titanium Project—$40 Million

5.8.1 Mining Business Case—Titanium Project—$40 Million

The project entails the financing the construction of a licensed titanium project after the feasibility study and testwork is completed (Table 5.14).

Table 5.14 Mining business case—titanium project—$40 million

<table>
<tr><th>#</th><th>Item</th><th>Description</th></tr>
<tr><td>1</td><td>Project stage</td><td>• Mine development: Opencast titanium mining project, permitted, feasibility study and significant testwork completed.
• Mine to be brought into production in phases.
• Seeking funds for construction.</td></tr>
<tr><td>2</td><td>Location</td><td>• Australasia</td></tr>
<tr><td>3</td><td>History</td><td>• Project has been developed over a 5-year period.</td></tr>
<tr><td>4</td><td>Corporate structure</td><td>• SPV
– 70% owned by Holdco
– 30% owned by local investor</td></tr>
<tr><td>5</td><td>Development team</td><td>• Mid-tier mining company</td></tr>
<tr><td>6</td><td>Permits and licences</td><td>• Mining licence secured for 30 years</td></tr>
<tr><td>7</td><td>Geology</td><td>• Mineralised zone comprising several massive titaniferous seams
• Average thickness: 90 m
• Mineralised zone strikes for 10 km and dips 75°
• Within the mineralised zone there are ten main seams, comprising massive titaniferous magnetite</td></tr>
<tr><td>8</td><td>Exploration</td><td>• Exploration work includes:
– 20 km of drilling
– Field mapping
– Trenching
– High-resolution aeromagnetic survey
– Seismic survey
– Aerial photography</td></tr>
<tr><td>9</td><td>Resources and reserves</td><td>JORC classification
Resource classification:
<table>
<tr><th></th><th>%Fe</th><th>%TiO_2</th><th>Mt</th></tr>
<tr><td>Resources</td><td>25</td><td>15</td><td>300</td></tr>
<tr><td>Reserves</td><td>25</td><td>15</td><td>50</td></tr>
</table></td></tr>
<tr><td>10</td><td>Mining</td><td>• Mining method: opencast mining
• Stripping ratio: 3:1
• Contractor mining</td></tr>
<tr><td>11</td><td>Processing</td><td>• Mineral concentrator process plant designed entailing:
– Primary and secondary crushing
– Dense medium separation
– Primary milling
– Low-intensity magnetic separation
– Magnetics dewatering and regrind
– Smelting, using a toll smelter</td></tr>
<tr><td>12</td><td>Production parameters</td><td>• Phased ramp-up
• Peak production: 2.5 Mtpa ore
• Stripping ratio: 3:1
• Life of mine: +25 years</td></tr>
<tr><td>13</td><td>Infrastructure</td><td>• Mineral concentrator
• Railway siding to be constructed
• Power plant 15 MW
• 50 km water pipeline
• Discard dump
• Mine offices, workshop, stores, change house</td></tr>
</table>

(continued)

Table 5.14 (continued)

#	Item	Description
		• Mine adits and waiting places • Telecommunication
14	Products	• The mine will produce: – Ilmenite concentrate – Titanium slag – Pig iron
15	Markets	• The product is sold to a metal's commodity trader – Ilmenite concentrate: \$180/t – Titanium slag: \$1000/t – Pig iron: \$200/t
16	Logistics	• Rail transport: • 180 ktpa of ilmenite concentrate transported via rail to smelter • Titanium slag and pig iron railed to port
17	Human resources	• The total workforce of the mine and mineral concentrator: 300 • All employees live off-site near the mine
18	Environmental	• EIA approved
19	Social and community	• Social and labour plan approved by community
20	Financial	• CAPEX: \$120 million • OPEX: \$14/t ROM • NPV: \$65 million • IRR: 22% • Payback period: 6 years
21	Project plan	• Construction: Month 1–14 • Operation: Month 15
22	Funding requirements	• \$ 40 million for mine development activities

Source: Authors' table

5.8.2 Mining Capital Term Sheet and Transaction Structure—Titanium Project—$40 Million

Term Sheet

The mining company pursued the *Corporate Bond Finance* route to finance the project. Corporate bonds were issued, valued each at $1,00,000 at 8% interest to raise the required CAPEX of $40 million for the implementation of the titanium project (Table 5.15).

Table 5.15 Term sheet development finance—titanium project—$40 million

Financial instrument	• Secured redeemable bonds
Corporation:	• Holdco
Holder	• Investor
Period	• 5 years
Interest	• 8% pa
Value per bond	• $1,00,000
Value of issued bonds	• $4,00,00,000
Trustee	• Acts as the administrator and paymaster
Terms	• The corporation promises the holder to pay on date X on presentation of the bond the sum of $100,00 plus interest at the office of the trustee • Bonds are issuable as fully registered bonds in denominations of $1,00,000 • All bonds rank equally and with priority of preference • The bonds will not be redeemable on or before date X • After date X, the bonds will be redeemable at the corporations sole option on 60 days' notice

Source: Authors' table based on typical corporate bond finance term sheet

Transaction Structure

Holdco issues corporate bond notes, which are bought by the investor, with the trustee acting as paymaster and administrator of the transaction (Fig. 5.8).

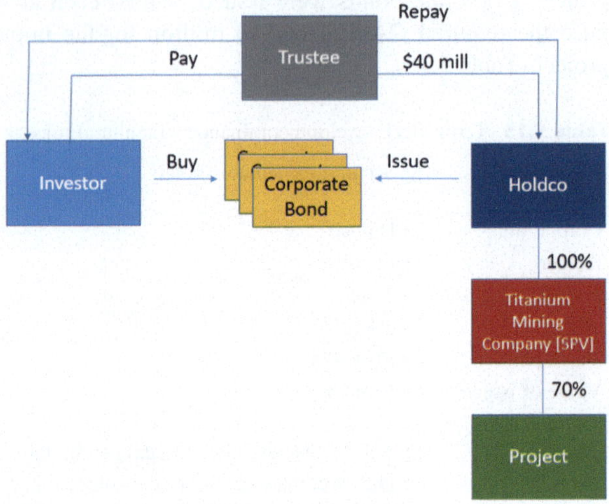

Fig. 5.8 Transaction structure—corporate bond finance—titanium project (Source: The author)

5.9 Streaming Finance—Gold Project—$75 Million

5.9.1 Mining Business Case—Gold Project—$75 Million

The project entails the financing of the construction of a North American underground gold mine development project [57, 58] (Table 5.16).

Table 5.16 Mining business case—gold project—$75 million

#	Item	Description
1	Project stage	• Advanced stage underground gold development project
2	Location	• North America
3	History	• Several exploration campaigns, and technical studies conducted over past decade
4	Corporate structure	• Holdco owns 100% of the project SPV • SPV owns 100% of the licences
5	Development team	• Mid-tier mining company
6	Permits and licences	• Several contiguous mining leases, patented claims, mining licences of occupation and a staked claim
7	Geology	• The gold project is located in one of North America's preeminent gold producing districts • Gold mineralisation is characterised by vein and sulphide replacement style mineralisation hosted within basalt units and felsic intrusive rock

(continued)

Table 5.16 (continued)

<table>
<tr><th>#</th><th>Item</th><th>Description</th></tr>
<tr><td>8</td><td>Exploration</td><td>• SPV conducted extensive exploration on project:
– Geological mapping
– Re-logging of selected historical boreholes
– Digital compilation of available historical data
– Ground and airborne magnetic surveys
– Trenching
– Channel sampling
– Airborne geophysical surveys
– Petrographic study
– Topographic survey
– Data modelling
– Several drilling programmes
• Drilling: +3,55,000 m drilled
• Resource estimate based on 820 boreholes drilled</td></tr>
<tr><td>9</td><td>Resources and reserves</td><td>Mineral resource statement
<table>
<tr><th>Category</th><th>Tons (Mt)</th><th>Grade (g/t)</th><th>Contained gold (Moz)</th></tr>
<tr><td>Measured and indicated</td><td>4.1</td><td>8.5</td><td>1.1</td></tr>
<tr><td>Inferred</td><td>7.5</td><td>9.3</td><td>2.2</td></tr>
</table></td></tr>
<tr><td>10</td><td>Mining</td><td>• Mining method: Underground trackless long hole open stoping, and cut and fill stoping operations</td></tr>
<tr><td>11</td><td>Mine equipment</td><td>• Mining equipment:
– LHDs
– Jumbo drills
– Haul trucks
– Locomotives
– Train cars
– 2387 trucks
– Service trucks</td></tr>
<tr><td>12</td><td>Processing</td><td>• Mineral processing plant:
– Semi-autogenous grinding mill
– Ball mill
– Hydrocyclones circuit
– Carbon-in-leach circuit
– Electrowinning and the gravity circuit both produce a high-grade gold concentrate
– Electric induction furnace to produce doré
– Cyanide destruction process</td></tr>
<tr><td>13</td><td>Production parameters</td><td>• LoM production tons: 9.13 M
• LoM recovered ounces 2.19 M
• Average production rate: 8,09,890 tpa
• Life of mine: 13 years
• Process recovery: 92.5%</td></tr>
</table>

(continued)

Table 5.16 (continued)

#	Item	Description	
14	Infrastructure	• The main surface infrastructure includes: – Hoist, head frame, and hoist house – Processing plant – Tailings management facility – Effluent treatment plant – Electric power supply and substation A – Propane storage tanks – Compressed air supply – Process and potable water supplies – Sewage works – Mine ventilation fans and heater house – Offices, shop, warehouse, core shack, and storage buildings – IT and communication technology	
15	Products	• Gold doré bars 99.9%	
16	Markets	• Gold streaming company	
17	Human resources	• The total workforce of the mine: +200 • All employees live off-site	
18	Environmental	• The project site is situated adjacent to a recreational lake. • Emphasis for physical environmental sensitivities has been placed on potential off-site discharges of water, dust and noise.	
19	Social	• Consultation with aboriginal communities under the guidance of government agencies.	
20	Financial	Key financial indicators	
		CAPEX	C$224 million
		OPEX	C$151/t
		NPV (5%)	C$531 million
		IRR	27%
		Payback period	3.7 years
21	Funding requirement	• $75 million for the construction of the gold mine	

Source: Authors' table based on Economic Assessment of the Phoenix Gold Project by SRK [57]

5.9.2 Mining Capital Term Sheet and Transaction Structure—Gold Project—$75 Million

Term Sheet

The mine developer selected *Streaming Finance* to finance for the implementation of the gold project (Table 5.17).

Table 5.17 Term sheet development finance—gold project—$75 million

Seller	• Mining company (SPV)
Purchaser	• Streaming company
Amount	• $75 million
Gold stream agreement	• Streaming company will provide mining company with a deposit of $75 million to be used for the construction of the gold project • Mining company agrees to deliver 6.3% of the projected annual gold production from the project until 1,35,000 ounces have been delivered and 3.15% thereafter • Streaming company purchase price per ounce will be 25% of the London spot price for one ounce of refined gold at the time of delivery
Obligations purchaser	• Purchaser will provide the provide mining company with a $75 million deposit • The deposit will be provided over five instalments: – $10 million on signing of the agreement – $20 million on closing – 3 × instalments of $15 million
Obligation by company	• The company will use the proceeds towards the construction of the gold project • Mining company agrees to deliver 6.3% of the annual projected gold production from the gold project to the streamer for up to 1,35,000 ounces, and 3.15% thereafter • Streamer will pay the company in cash 25% of the spot price for each delivered ounce
Term	• The term of this agreement is 40 years • The term shall be extendable at either party's option for successive 10-year periods
Covenants	• Conduct of operations: All decisions regarding the development and operation of the project shall be made by the seller in its sole discretion • Seller and purchaser shall always maintain their corporate existence • The seller shall not process other minerals through the facilities • The seller shall maintain insurance for the gold project • The purchaser has security interests in the assets of the project • If the seller arranges senior financing that is to be secured by any senior financing party, the purchaser will enter into a subordination agreement with the relevant senior financing party • The seller hereby agrees that it shall not sell unprocessed ore • The seller shall complete a minimum 38,000 m drilling programme

Source: Authors' table based on Streaming Agreement Rubicon Minerals Corporation with Royal Gold Inc. [58]

Transaction Structure

The transaction structure displays SPV receiving a loan of $75 million in exchange for delivering 6.3% of the gold output sold at 25% of the London spot price of the current gold price until 1,35,000 ounces have been delivered (Fig. 5.9).

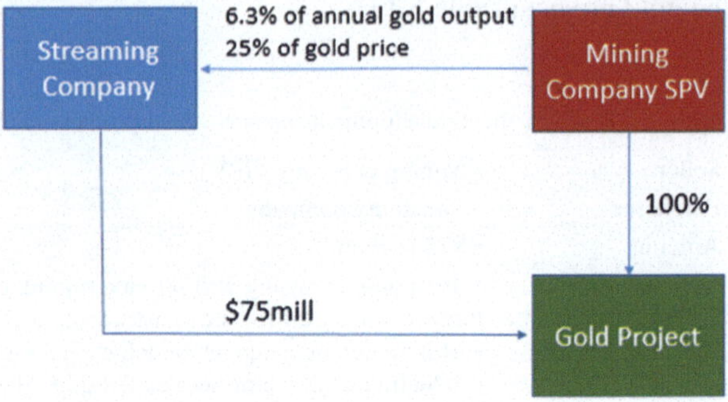

Fig. 5.9 Transaction structure—streaming finance—gold project (Source: The author)

5.10 Royalty Finance—Gold–Silver Project—$10 Million

5.10.1 Mining Business Case—Gold–Silver Project—$10 Million

The project entails the start-up of a gold–silver heap leaching project in North America [59, 60] (Table 5.18).

Table 5.18 Mining business case—gold project—$10 million

#	Item	Description
1	Project stage	• Advanced stage gold and silver development project
2	Location	• North America
3	History	• Previous mining at the property included the construction of open pit excavations, a waste rock dump and a heap leach pad • 1968–1982: Drilling of +30,000 m
4	Corporate structure	• Holdco acquired 80% of project SPV • SPV owns 100% of leasehold interest in a mining lease
5	Development team	• Junior mining company
6	Permits and licences	• Patented mineral claims and unpatented federal mining claims
7	Geology	• The project is located in a gold district • Geology is characterised by a polymetallic skarn overprinted by late-stage epithermal gold mineralisation concentrated along two shallowly dipping faults • Weathering and oxidation caused formation of oxide mineralisation from which gold is recoverable by cyanide heap leaching
8	Exploration	• SPV conducted extensive exploration on project

(continued)

Table 5.18 (continued)

<table>
<tr><th>#</th><th>Item</th><th colspan="4">Description</th></tr>
<tr><td rowspan="5">9</td><td rowspan="5">Resources and reserves</td><td colspan="4">Reserve statement</td></tr>
<tr><td></td><td>Tons (000)</td><td>Au (g/t)</td><td>Ag (g/t)</td></tr>
<tr><td>Proven</td><td>923</td><td>0.032</td><td>0.155</td></tr>
<tr><td>Probable</td><td>21,604</td><td>0.021</td><td>0.134</td></tr>
<tr><td>Total</td><td>22,527</td><td>0.022</td><td>0.134</td></tr>
<tr><td>10</td><td>Mining</td><td colspan="4">• Mining method: open pit mining method
• Stripping ratio 2.5:1</td></tr>
<tr><td>11</td><td>Mine equipment</td><td colspan="4">• Equipment:
– Blast hole drills
– Hydraulic shovel
– Rubber-tired wheel loaders
– Off-highway 100-ton trucks
– Graders
– Track dozers
– Water truck</td></tr>
<tr><td>12</td><td>Processing</td><td colspan="4">• Heap leaching operation: 22.5 Mt
• Crushing: Primary and secondary
• Crushed ore stacked on heap leach pad
• Ore leach cycle: 210 days
• Ore stacking rate: 550 tph
• Average gold recovery: 80%
• Average silver recovery: 30%
• Heap leach pad and associated facilities: 150 acres</td></tr>
<tr><td>13</td><td>Production parameters</td><td colspan="4">• Production rate: 10,000 t/d
• Life of mine: 7 years
• Payable metal:
– Gold: 415 koz
– Silver: 1690 koz
• Recoveries:
– Gold: 76%
– Silver: 39%
• ROM grades:
– Gold: 0.024 oz/t
– Silver: 0.197 oz/t
• Average annual production:
– Gold: 68,600 oz/annum
– Silver: 2,79,400 oz/annum</td></tr>
<tr><td>14</td><td>Infrastructure</td><td colspan="4">• The nearest power line 17 miles from the project site.
• Current power: on-site diesel generated electrical power
• Site access: Highway and gravel-surfaced public and private roads.</td></tr>
<tr><td>15</td><td>Products</td><td colspan="4">• Product: gold and silver doré
• Au price: $1300/oz
• Ag price: $20.00/oz</td></tr>
<tr><td>16</td><td>Markets</td><td colspan="4">• Prices:
– Gold price: $1300/oz
– Silver price: $20.00/oz
• Gold royalty company</td></tr>
</table>

(continued)

Table 5.18 (continued)

#	Item	Description	
17	Logistics	• Site access: Highway and gravel-surfaced public and private roads	
18	Human resources	• All employees live off-site • Four crews operating on 12 h rotating shifts. • Crews will be transported to the site in company supplied vans.	
19	Environmental	• All buildings have been removed and the leach pad associated with previous mining has been covered with soil, recontoured, and seeded. • SPV has no environmental liabilities related to this previous mining activity.	
20	Financial	Key financial indicators	
		CAPEX	Initial $91.7 million
		OPEX	$11.5/t ore
		NPV (8%)	$61 million
		IRR	26%
		Payback period	3 years
21	Funding requirement	• Total CAPEX requirement $91 million for the construction of the gold mine • Seeking part funding of $10 mil from royalty finance	

Source: Authors' table based on summary of NI 43–101 Technical Report Feasibility Study Mt. Hamilton Gold and Silver Project SRK Consulting [59]

5.10.2 Mining Capital Term Sheet and Transaction Structure—Gold Project—$10 Million

Term Sheet

The mine developer secured *Royalty Finance*, type Net Smelter Return Royalty, to secure $10 million, as part of the total financing package (Table 5.19).

Table 5.19 Term sheet royalty finance—gold project—$10 million

Owner	• 100% undivided interest in certain unpatented mining claims seeks to grant to the royalty holder a certain net smelter returns royalty.
Covenantor	• The covenantor provides a guarantee to and in favour of the royalty holder.
Production	• The quantity of refined gold and silver that is produced to the owner's account by a refinery during a month.
Monthly average gold price	• Average London Bullion Market Association: Gold Fix
Monthly average silver price	• Average London Bullion Market Association: Silver Fix
Net smelter returns	• Gross proceeds for such month from the sale of such product less allowable deductions for such month.
Products	• All gold or silver bearing ores mined from the property and all concentrates and other mineral products, metals or minerals which are derived therefrom.
Investment/payment	• US$10 million
Net smelter return royalty	• In consideration of the payment of the sum of US$10 million which shall be paid by the royalty holder to the owner, the owner agrees to pay to the royalty holder the royalty in respect of products.
Royalty	• 2.4% of the gross value of recoverable minerals contained in such products.
Term	• The entire term of the mining lease
Covenant	• The royalty will be running with the owned property
Tailings and residue	• All tailing, residues, waste rock, spoiled leach materials, and other materials are subject to the obligation to pay the royalty
Purchase of royalty	• US$12.0 million
Warranties	• The owner warrants to and in favour of the royalty holder that it has the corporate company power, capacity and authority to execute, deliver and perform this agreement.
Transfer rights of the royalty holder	• The royalty holder shall have the right to transfer its rights and obligations under this agreement to one or more lenders providing financing to the royalty holder.
Owners discretion	• The owner will have complete discretion concerning the nature, timing and extent of all exploration, development, mining and other operations conducted on or for the benefit of the property.

Source: Authors' table based on Net Smelter Return Agreement Solitario Exploration & Royalty Corp, Mt. Hamilton LLC and Sandstorm Gold [60]

Transaction Structure

The transaction structure displays the royalty company providing a $10 million finance facility to the SPV in exchange for a royalty of 2.4% of the gross value of recoverable minerals (Fig. 5.10).

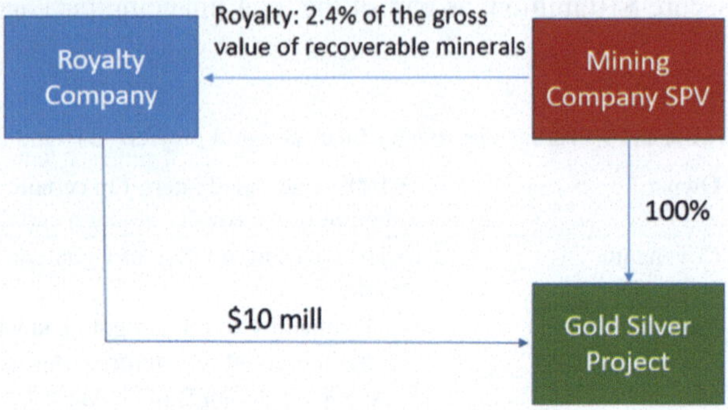

Fig. 5.10 Transaction structure—royalty finance—gold project (Source: The author)

5.11 Offtake Finance—Coal Mine—$10 Million

5.11.1 Mining Business Case—Coal Mine—$10 Million

The project entails an operating coal mine seeking CAPEX to fund operational costs and expand production (Table 5.20).

Table 5.20 Mining business case—coal mine—$10 million

#	Item	Description
1	Project stage	• Operating coal mine • Requires CAPEX to fund operational costs and expand production
2	Location	• Asia
3	Corporate structure	• Coal mining company SPV concluded an offtake funding agreement with an international commodity trader
4	Licence	• The coal mine is fully licenced: Mining licence
5	Geology	• Coalfield xxx
6	Exploration	• Exploration work conducted: Geological mapping, geophysical investigations, geotechnical tests and drilling. • A structured exploration programme is required to delineate coal blocks for the next 5 years.
7	Resources and reserves	• Reserves: 10 million tons • Resources: 25 million tons
8	Mining	• Mining method: underground board and pillar mining • Recapitalisation entails the introduction of new mechanised underground mining equipment. • The mine layout needs to be adapted for mechanisation.

(continued)

Table 5.20 (continued)

#	Item	Description
9	Processing	• The high-grade coal is crushed and screened to produce nuts, peas and duff. • The current crusher is poorly maintained and produces excess duff which sells at a discount. • Current plant capacity: 25,000 tph, availability 30% • Planned capacity: 1,00,000 tpm, availability 85%
10	Production parameters	• Production rate actual: 25,000 tpm ROM. • Production rate planned: 1,00,000 tpm • Life of mine: +10 years
11	Infrastructure	• Mine offices • A weighbridge • A crusher • Access road • Power supply: 1 MW • Conveyor belts • Mine adits and waiting places • Water supply via pump station at local river • Telecommunication
12	Products	• Coal nuts, peas and duff
13	Markets	• Regional boiler industry • Cement factory
14	Logistics	• The coal is sold ex gate free on trucks (FOT)
15	Human resources	• The total workforce of the mine: 200 • All employees live off-site near the mine
16	Financial	• CAPEX required to recapitalise the mine: $10 million for new equipment, reconfiguration of the underground mining sections. • Financial indicators: NPV: $20 mill, IRR 30%, payback period 3 years.
17	Funding requirements	• $10 million to recapitalise the mining operation and ramp up production.

Source: Authors' table

5.11.2 Mining Capital Term Sheet and Transaction Structure—Coal Mine—$10 Million

Transaction Structure

The mine produced an attractive coal export product, which attracted an international coal commodity trader to provide *Offtake Finance*. The capital injection enabled the recapitalisation of the mine. In exchange, the rights to all the coal products were passed on to the commodity trader, at a discounted price (Table 5.21).

Table 5.21 Term sheet offtake finance—coal mine—$10 million

Seller	• Coal mining company
Buyer	• Commodity trader
Prepayment	• $10 million
Product	• Coal • Quantity: 1,00,000 tons per month – Quality: – Calorific value 24 MJ/kg, – 24% ash – 20% volatiles – 1% Sulphur – Sizing: –50 mm
Period	• 36 months
Start date	• Date X
Price	• Coal index based price, less 18% to account for upfront investment
Delivery point	• Ex mine gate
Prepayment	• 100% prepayment to be made by buyer, 20 business days prior to the commencement of the month that coal will be delivered to delivery point

Source: Authors' table

Transaction Structure

The international coal commodity trader provided a prepayment of $10 million to the SPV in exchange in for a delivery commitment for 1,00,000 tpm at a discount of 18% to the prevalent global coal index for 36 months (Fig. 5.11).

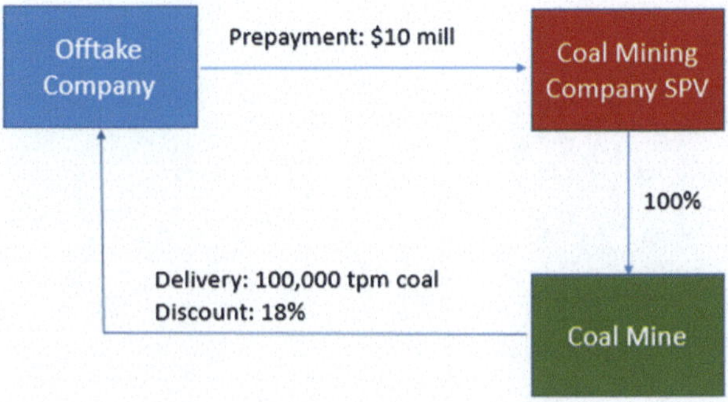

Fig. 5.11 Transaction structure—offtake finance—gold project (Source: The author)

5.12 Insurance Wrapped Finance—Anthracite Mine—$11 Million

5.12.1 Mining Business Case—Anthracite Mine—$11 Million

The project entails an operating anthracite mine seeking $11 million to expand production to meet growing market demands (Table 5.22).

Table 5.22 Mining business case—anthracite mine—$11 million

#	Item	Description
1	Project stage	• Operational anthracite mine seeking funds to expand the business. • The mine cannot meet the growing market demands.
2	Location	• South America
3	Legal	• SPV: – 50% owned by shareholder 1 – 50% owned by shareholder 2
4	Licence	• Mining licence issued to SPV for 15 years
5	Geology	• The geological setting is discrete, faulted high-grade anthracite blocks • The anthracite seam is 2 m thick
6	Exploration	• Exploration work conducted: geological mapping, geophysical investigations, geotechnical tests and drilling.
7	Resources and reserves	• Reserves: 5 million tons • Resources: 15 million tons
8	Mining	• Mining method: small-scale underground board and pillar mining • Recapitalisation entails the introduction of modern mechanised underground mining equipment. • The mine layout needs to be adapted to be mechanised.
9	Processing	• Crushing and screening
10	Production parameters	• Production rate actual: 5000 tpm ROM • Production rate planned: 20,000 tpm • Life of mine: +10 years
11	Infrastructure	• Mine offices • A weighbridge • A crusher • Access road • Power supply: 0.8 MW • Conveyor belts • Mine adits and waiting places • Water supply via pump station at local river • Telecommunication
12	Equipment	• The current mine equipment consists of: – Front end loaders × 2 – Articulated dump trucks × 2 – Compressors × 4 – Generators × 4 – Mine vehicles × 5 • 80% of the equipment is in poor condition and needs to be refurbished or replaced to enable the production plan to be achieved.

(continued)

Table 5.22 (continued)

#	Item	Description
13	Products	• Anthracite 0–50 mm for ferrochrome industry
14	Markets	• Global community traders seeking product for ferrochrome industry
15	Logistics	• The anthracite is sold free on truck (FOT).
16	Human resources	• The total workforce of the mine: 200 • All employees live off-site near the mine
17	Financial	• CAPEX required to recapitalise the mine: $11 million for new equipment, reconfiguration of the underground mining sections. • Financial indicators: NPV: $15 million, IRR 25%, payback period 3 years.
18	Funding requirements	• $11 million to recapitalise the mining operation and ramp up production.

Source: Authors' table

5.12.2 Mining Capital Term Sheet and Transaction Structure—Anthracite Mine—$11 Million

Term Sheet

The mine owners decided to opt for *Insurance Wrapped Finance* given the presence of an operation and market for the product. Insurance backed funding entails an insurer backing the business case and insuring the funder of the project, i.e. the banks. The banks pay the supplier of equipment and mining services to recapitalise the mine (Table 5.23).

Table 5.23 Term sheet offtake finance—anthracite mine—$11 million

Borrower	• SPV: Provides business case to support a recapitalisation of $11 million. • The bulk of the capital is for underground mining equipment and contracting services.
Insurance company	• International insurer approves the business case and insures the bank against a default of the business case.
Bank	• A commercial bank provides a debt facility to recapitalise mine. • The mine equipment supplier and mine service providers are paid by the bank for equipment and services to be rendered to recapitalise the mine.
Equipment suppliers	• Equipment supplier 1 provides underground mining equipment • Equipment supplier 2 provides mining surface equipment
Mining services provider	• The mining services providers include: – Drilling company – Mine planning company – Mining contractor

(continued)

Table 5.23 (continued)

Requirements by bank and insurer	• A mining business case • Mining licence • Mineral resources and reserves, signed off by a competent person • A mine plan • A mine development and operations team • A market for the anthracite product
Facilities	• Total value of services: $11 million • Value of mining equipment: $8 million • Value of services: $3 million
Interest rate	• 12% pa
Term	• 5 years
Security	• Mining licence of SPV

Source: Authors' table

Transaction Structure

The transaction structure outlines how the insurer approves the business case of the SPV and insures the bank. The bank in turn finances equipment $8 million and mining services $3 million. The SPV repays the bank loan, $11 million plus interest (Fig. 5.12).

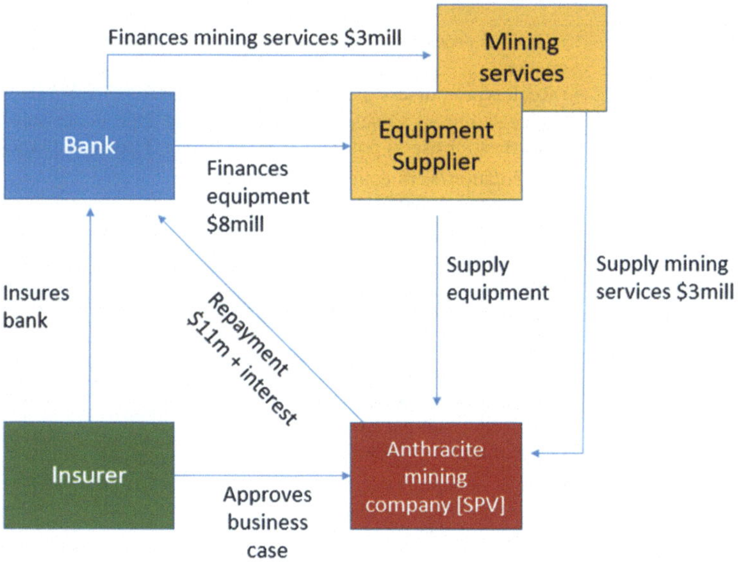

Fig. 5.12 Transaction structure—insurance wrapped finance—anthracite mine (Source: The author)

5.13 Contractor Vendor Finance—Dolomite Mine—$7 Million

5.13.1 Mining Business Case—Coal Mine—Dolomite Mine—$7 Million

An operational dolomite mine seeks capital to ramp up production, by acquiring additional opencast mining and processing equipment valued at $7 million (Table 5.24).

Table 5.24 Mining business case—dolomite mine—$7 million

#	Item	Description
1	Project stage	• Operational dolomite quarrying business seeking funds to expand business to meet growing market demand. • Current operation: Medium operation, 160 tph • Planned: Ramp-up of operations. Require additional mining equipment and installation of new 550 tph crushing and screening plant.
2	History	• The quarry has been operational for +15 years
3	Location	• Europe • 10 km from nearest town
4	Corporate structure	• SPV: 100% owned by family business
5	Permits and licences	• Mining licence issued to SPV for 20 years
6	Geology	• The main rock units occurring within the property are porphyritic granites. • Rock types possess the appropriate aggregate crushing values aggregate impact values, specific gravity and appropriate polishing properties for use as polished slabs and other decorative purposes.
7	Exploration	• Exploration work conducted to date: – Geological mapping – Geophysical investigation – Geotechnical tests – Drilling
8	Resources and reserves	• Mineable reserve: 40 million tons
9	Mining	• Opencast mining • Production rate actual: 80,000 tpm • Production rate planned: 2,80,000 tpm • Life of mine: 12 years
10	Mine equipment	• Current equipment: – 160 tph crushing and screening plant – Excavator 45 t × 2 – Dozer D8 × 1 – Articulated dump trucks 30 t × 3 – Front end loaders 966 × 3 – Drillrig × 1 – Grader × 1 – Crane truck × 1

(continued)

Table 5.24 (continued)

#	Item	Description
		– Service truck × 1 – Ambulance × 1 – Mobile workshop × 1 – Weighbridge × 1 x – Diesel truck × 1 • Planned equipment: – Articulated dump trucks 40 t × 8 – Excavator 80 t × 2
11	Processing	• Crushing and screening plant • Current plant capacity: 160 tph • Planned capacity: 550 tpm
12	Production parameters	• Mineable reserves: 40 Mt • ROM/annum: 1 Mtpa
13	Infrastructure	• Mine is established with mineral processing plant, mine offices, workshop, mine power, security access.
14	Products	• Crushed stones • Free market slates • Material for sand filling and piling • Quality: Dolomite, silica content 4–5%
14	Markets	• Construction industry for roads, housing estates, buildings, bridges for urban cities. • Several expression of interest secured to purchase product
15	Logistics	• Product sold free on truck (FOT) ex mine gate • Product loaded free on truck onto 30-ton trailer tippers
16	Human resources	• The total workforce of the mine: 120 • All employees live off-site
15	Financial	• CAPEX required: $7 million • Profit after tax: $1 million pa • Liabilities: $1.5 million
17	Funding requirements	• Financing of $7 million to expand operations

Source: Authors' table

5.13.2 Mining Capital Term Sheet and Transaction Structure—Dolomite Mine—$7 Million

Term Sheet

The mining company (SPV) was seeking mining and processing equipment and services to ramp up production. The company selected a *Contractor Vendor Finance* solution as opposed to buying fleet directly. The SPV entered into a long-term contract with a mining contractor under a Build + Own + Operate Contract. Production could be ramped up to meet the growing market demand (Table 5.25).

Table 5.25 Term sheet contractor vendor finance—dolomite mine—$7 million

Funder	• Mining contracting company
Borrower	• SPV
Contract type	• Build, own, operate (BOO)
Scope of work	• Mining contracting company to provide: – Site establishment fees – Mining equipment – Mineral processing equipment – Operational management team • Required production rate: 1 Mtpa • Product: Sized crushed product
Requirements by borrower	• Borrower to ensure: – Mining licence is intact – Geological model is available – Market for product guaranteed with short-, medium- and long-term contracts
Facilities	• Value of mining equipment: $5.5 million • Site establishment fee: $1.5 million
Commercial rates	• SPV to pay contractor: • Fixed fee: $6,00,000 per month • Variable fee: $15/ton
Term	• 5 years
Security	• Mining licence

Source: Authors' table

Transaction Structure

The dolomite mine SPV entered into a Build + Own + Operate contract with a mining contractor for mining services. The mining equipment to the value of $7 million was financed by the mining contractor in exchange for a 5-year contract for specified 83,000 tpm volume (Fig. 5.13).

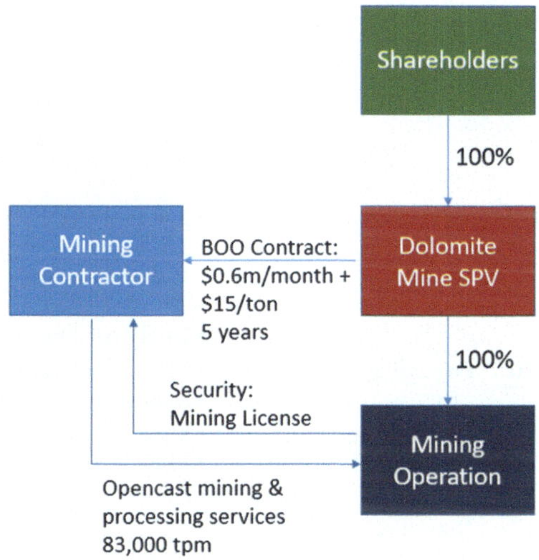

Fig. 5.13 Transaction structure—contractor vendor finance—dolomite mine (Source: The author)

5.14 Equipment Finance—Mining Contractor—$0.6 Million

5.14.1 Mining Business Case—Mining Contractor—$0.6 Million

A mining contractor is seeking to purchase opencast mining equipment after having secured a 3-year mining contract to mine manganese ore for a large mining company (Table 5.26).

Table 5.26 Mining business case—mining contractor—$0.6 million

#	Item	Description
1	Project stage	• Operation • Mining contractor seeking to equipment finance to purchase 2 × articulated dump trucks
2	Location	• Southern Africa
3	Legal	• Mining contractor has secured a 3-year mining contract to mine 4.5 Mt of manganese for a large mining company.
4	Licence	• The mining licence is held by the mining company.
5	Geology	• Hardrock manganese deposit located in dolomite setting.
6	Exploration	• The client, a mining company, has concluded a drilling campaign and developed a geological model. • The mining contractor is required to develop a mine plan.

(continued)

Table 5.26 (continued)

#	Item	Description
7	Resources and reserves	• Mineable reserves allocated to mining contract: 4.5 Mt
8	Mining	• Mining method: opencast, truck and shovel • Scope of work: Full opencast mining services • Stripping ratio: 2:1 • Production parameters: 1.5 Mtpa ROM • Life of contract: 3 years
9	Equipment	• Mining fleet: • 3 × 85 ton excavators • 12 × 40 ton ADT trucks • 1 × drilling machines • 2 × dozers D9 • 1 × grader • 2 × front end loader • 1 × explosive truck • 1 × water tanker • 1 × diesel bowser • Utility equipment: LDVs, service vehicle, crane, mobile lights
10	Processing	• Mining contractor to provide modular crushing plant. • Manganese ore to be crushed to −50 mm
11	Infrastructure	• Mine company provides water and power to mining contractor. • Mining contractor to site establish and erect workshop, stores, mobile office, change house.
12	Human resources	• The total workforce of the mining contractor: 150
13	Financial	Key financial indicators Term of contract: 3 years Mining cost: $7/ton ROM Value of contract: $32 million
14	Funding requirements	Funding to purchase 2 × articulated dump trucks: $0.6 million

Source: Authors' table

5.14.2 Mining Capital Term Sheet and Transaction—Mining Contractor—$0.6 Million

Term Sheet

The mining contractor selected *Equipment Finance* to finance the mining equipment. Three proposals from commercial banks were evaluated before one commercial bank was selected (Table 5.27).

Table 5.27 Term sheet equipment finance—mining contractor—$0.6 million

Lender	• Commercial bank
Borrower	• Mining contractor
Type of agreement:	• Instalment sale agreement and credit facility
Description of goods	• 2 × 40-ton articulated dump trucks supplied by equipment supplier
Cash value	• $4,90,714
VAT amount	• $68,706
Total cash value	• $5,59,420
Interest	• 15.5%
$ value interest	• $1,57,435
Total amount repayable	• $6,05,009
Suretyship	• Shareholders of mining contractor
Number of instalments	• 48
Amount of each instalment	• $12,604

Source: Authors' table based on typical equipment finance term sheet from commercial banks

Transaction Structure

The mining contractor SPV secures equipment finance, $0.6 million, from the commercial bank against a contract of $31.5 million where it provides opencast mining services to the manganese mine. The SPV's shareholders provide surety to the commercial bank. The equipment finance facility for 2 × ADTs is paid back over 48 months (Fig. 5.14).

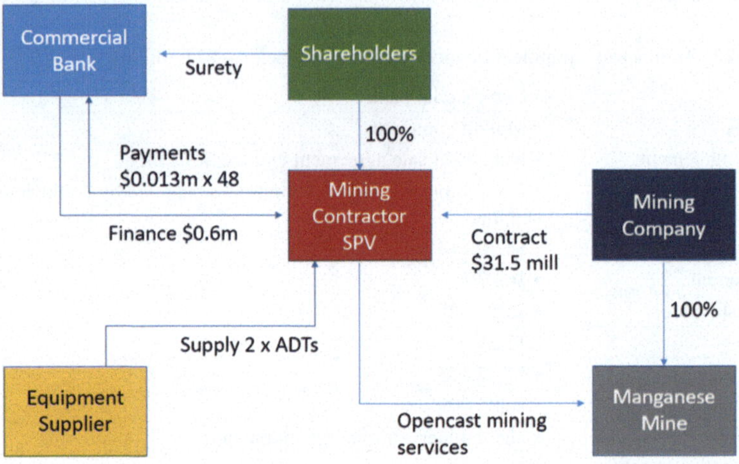

Fig. 5.14 Transaction structure—equipment finance—mining contractor (Source: The author)

5.15 Disinvestment Finance—Iron Ore Operation—$6 Million

5.15.1 Mining Business Case—Iron Ore Operation—$6 Million

An ageing iron ore operation, having run out of large reserves, is seeking finance to settle debt and reengineer the business to an iron ore logistics hub (Table 5.28).

Table 5.28 Mining business case—iron ore operation—$6 million

#	Item	Description
1	Project stage	• Mine closure • Iron ore operation entailing mineral processing plant, a railway siding and a remnant iron ore reserve. • Seeking investor to purchase the remaining remnant iron ore reserve of 10 Mt.
2	Location	• South America
3	History	• The iron ore company ran out of large economic iron ore reserves and became unprofitable. • The business will be reengineered to become an iron ore logistics hub for neighbouring iron ore mines. • The business requires capital to settle debt.

(continued)

Table 5.28 (continued)

<table>
<tr><th>#</th><th>Item</th><th colspan="3">Description</th></tr>
<tr><td>4</td><td>Corporate structure</td><td colspan="3">• Holdco holds 100% shareholding in 2 × SPVs
• SPV 1 owns the mineral processing complex and the siding
• SPV 2 houses the remnant iron ore deposit.</td></tr>
<tr><td>5</td><td>Permits and licences</td><td colspan="3">• The iron ore operation is fully licenced</td></tr>
<tr><td>6</td><td>Geology</td><td colspan="3">• The iron ore mining operation is located in a traditional iron ore region.
• The iron ore field has become fragmented, as major mining companies have mined the major reserves.
• Junior mining companies are mining the remnant iron ore deposits and seeking further iron ore deposits to mine.</td></tr>
<tr><td>6</td><td>Exploration</td><td colspan="3">• Extensive drilling conducted on remnant reserve.</td></tr>
<tr><td rowspan="3">7</td><td rowspan="3">Resources and reserves</td><td colspan="3">Resource classification of remnant deposit:</td></tr>
<tr><td></td><td>Fe%</td><td>Mill tons</td></tr>
<tr><td>Reserves</td><td>55</td><td>10</td></tr>
<tr><td>8</td><td>Mining</td><td colspan="3">• Mining method: opencast mining
• Contractor mining
• Life of mine of remnant reserve: +3 years</td></tr>
<tr><td>9</td><td>Processing</td><td colspan="3">• The mineral processing plant can be used to process external iron ore delivered to the iron ore complex.
• Processing circuit: Crushing, dense medium cyclone plant, spiral plant, thickener, high-grade screening</td></tr>
<tr><td>10</td><td>Infrastructure</td><td colspan="3">• Mineral processing plant
• Railways siding
• Workshop
• Pollution control dams
• Weighbridges
• Mine offices, workshop, stores, change house</td></tr>
<tr><td>11</td><td>Product</td><td colspan="3">• Iron concentrate: 62% Fe</td></tr>
<tr><td>12</td><td>Market</td><td colspan="3">• The mine produces iron ore for international commodity traders</td></tr>
<tr><td>13</td><td>Logistics</td><td colspan="3">• The railway siding links to export harbours</td></tr>
<tr><td>14</td><td>Human resources</td><td colspan="3">• The total workforce of the iron ore logistics complex: 150</td></tr>
<tr><td>15</td><td>Environmental</td><td colspan="3">• The iron ore complex has an approved EMP</td></tr>
<tr><td>16</td><td>Social</td><td colspan="3">• The buyer of the iron ore project needs to subscribe to the approved social and labour plan.
• As a condition of the sale, most of the staff and workforce need to come from the local community.</td></tr>
<tr><td>17</td><td>Financial</td><td colspan="3">Financials of remnant iron ore reserve:
• CAPEX: $5 million
• OPEX: $19/t ROM
• NPV: $8 million
• IRR: 20%
• Payback period: 2 years</td></tr>
<tr><td>18</td><td>Funding requirements</td><td colspan="3">• Holdco requires $6 million to settle debt and reengineer the business to an iron ore processing and logistics hub.</td></tr>
</table>

Source: Authors' table

5.15.2 Mining Capital Term Sheet and Transaction Case—Iron Ore Operation—$6 Million

Term Sheet

The iron ore mining company (Holdco) decided to sell the remaining iron ore reserve to a junior mining company—*Disinvestment Finance*—to raise funds to settle debt and reengineer the business to an iron ore processing and logistics hub (Table 5.29).

Table 5.29 Term sheet equipment finance—iron ore operation—$6 million

Investor	• Junior mining company
Seller	• Holdco
Transaction	• Investor will acquire 100% of SPV 2, including and any surface rights, servitudes or road access held by Holdco which are required to develop such mineral rights.
Consideration	• A cash payment of $1 million plus a royalty of $1 per metric ton of iron ore produced and sold from any of the mineral rights included in this agreement until maximum cash consideration of $6 million has been paid to Holdco.
Warrants	• Holdco warrants that it shall cause no harm or liability to the mineral or surface rights.
Condition precedent	• Completion of a due diligence process by investor which shall include drilling further holes on the mineral rights included in the transaction to expedite the mining and extraction process.
Approval	• This offer is subject to the unanimous approval of the transaction by the board of investor and Holdco.

Source: Authors' table

Transaction Structure

The transaction structure depicts the investor, the junior mining company, acquiring SPV2, which holds the remnant iron ore reserve, for cash and royalties totalling $6 million from Holdco (Fig. 5.15).

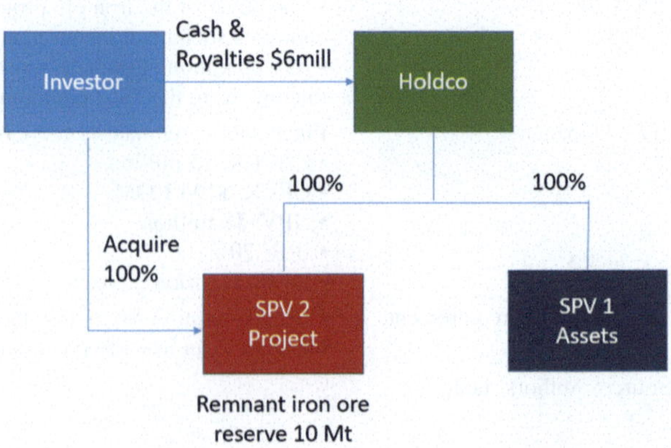

Fig. 5.15 Transaction structure—disinvestment finance—iron ore operation (Source: The author)

References

1. *The Northern Miner*. (2017, January). New frontiers in mining finance – Research report.
2. *Deloitte*. Tracking the trends 2018 – The top 10 issues mining companies will face in the coming year.
3. Notes from *Junior Indaba*, Johannesburg. (2019, June).
4. Cambridge House International. *Raw materials that fuel the green revolution*. Vancouver Resource Investment Conference 2018.
5. *Mining Journal*. The Global Mining Finance Guide 2014.
6. *E&Y*. Optimize for today? Build for tomorrow? Mergers, acquisitions and capital raising in mining and metals – 2018 outlook.
7. Sanderson, H. (2018, August). Coal industry between a rock and a hard place. *Financial Times*.
8. Mathews, C. (2018, June 6). *Era of mega-projects is over as lenders detail investment criteria*. https://www.miningmx.com/news/markets/33135-era-mega-projects-lenders-spell-investment-criteria/
9. Greve, N. (2018, February). *Juniors asked to consider royalty, streaming project finance deals*. https://www.miningmx.com/news/energy/31512-junior-asked-consider-royalty-streaming-project-finance-deals/
10. Solomons, S.-L. (2018, August). South Africa: Investing in junior mining is a game of patience. *Mining Review Africa*. https://www.miningreview.com/news/south-africa-investing-junior-mining-game-patience/
11. James, N. (2019, June). Strategic partnerships with midtiers, majors key to securing funding for juniors. *Junior Indaba*. https://www.engineeringnews.co.za/print-version/strategic-partnerships-with-midtiers-majors-key-to-securing-funding-for-juniors-2019-06-05
12. http://www.miningweekly.com/article/peabody-announces-400m-alabama-coal-acquisition-2018-09-21
13. http://www.miningweekly.com/article/barrick-gold-to-buy-randgold-resources-in-183bn-deal-2018-09-24
14. http://www.miningweekly.com/article/universals-a186m-deal-advances-2018-08-08
15. https://www.miningnews.net/capital-markets/news/1352159/artemis-finds-funding-for-paterson-province-exploration
16. https://www.miningweekly.com/article/millennium-secures-finance-for-nullagine-expansion-2018-06-06/rep_id:3650
17. https://www.miningweekly.com/article/nickel-mines-finds-funding-amends-ranger-deal-2019-05-21
18. http://www.miningweekly.com/article/lion-one-secures-funding-for-gold-project-2018-06-06
19. http://www.miningweekly.com/article/financing-of-ztk-transmission-line-to-benefit-rukwa-project-edenville-2018-06-22
20. http://www.miningweekly.com/article/citic-loans-ivanhoe-100m-2018-08-10
21. http://www.miningweekly.com/article/ncondezi-to-restructure-loan-2018-08-10
22. https://www.miningweekly.com/article/lucapa-finalises-idc-funding-for-mothae-2018-12-21

M. Seeger, *Mining Capital*, https://doi.org/10.1007/978-3-030-31225-1

23. http://www.mining.com/quebec-government-invests-1-2-million-advance-newmont-goldcorps-mine-future/
24. http://www.miningweekly.com/article/emmerson-raises-cash-for-moroccan-potash-project-2018-06-04
25. http://www.miningweekly.com/article/poseidon-unveils-a75m-capital-raising-2018-08-24
26. http://www.mining.com/mineral-resources-inks-1-15b-lithium-deal-albemarle/
27. https://www.miningweekly.com/article/teck-invests-in-british-columbias-stardust-project-2018-12-21
28. https://www.miningweekly.com/article/centaurus-sells-conquista-2018-11-23
29. https://www.miningweekly.com/article/centaurus-sells-conquista-2018-11-23
30. https://www.miningweekly.com/article/mustang-amends-arena-financing-2018-05-22
31. http://www.miningweekly.com/article/talga-raises-cash-for-product-developments-2018-06-28
32. https://www.miningweekly.com/article/peel-raises-cash-for-cobar-deposits-2019-06-06
33. https://www.miningweekly.com/article/galilee-raises-cash-for-glenaras-project-2018-07-10
34. https://www.miningnews.net/project-finance/news/1363346/venture-raising-to-fund-riley-revival
35. http://www.miningweekly.com/article/chinas-ganfeng-buys-sqms-stake-in-lithium-project-for-875m-2018-08-14
36. https://www.miningweekly.com/article/titan-raises-a11m-for-mirador-acquisition-2018-05-22
37. https://www.miningweekly.com/article/stonewall-raises-a6m-for-theta-hill-study-2018-11-12
38. https://www.miningweekly.com/article/fortescue-approves-287m-queens-valley-development-2019-05-22
39. https://m.miningweekly.com/article/new-botswana-coal-mine-development-under-way-minergy-2018-09-18
40. http://www.miningweekly.com/article/peabody-announces-400m-alabama-coal-acquisition-2018-09-21
41. http://www.miningweekly.com/article/thiess-wins-a480m-contract-for-qcoal-2018-06-29
42. https://www.miningweekly.com/article/bauba-signs-chrome-offtake-agreement-2018-08-08
43. https://www.miningweekly.com/article/lonmin-undertakes-refinancing-delivers-on-results-promise-2018-10-22
44. Seeger, M. (2007). *Development of a strategic and tactical game plan for junior mining companies*. PhD thesis, University of the Witwatersrand, South Africa.
45. Caselli, S., & Gatti, S. (2017). *Structured finance: Techniques, products and market* (2nd ed.). Cham: Springer.
46. Course Notes. (2005, August). *Wits CEE course: Financing of mining projects*. Course held at the University of the Witwatersrand.
47. https://mrmr.cim.org/en/standards/valuation-guidelines-for-mineral-properties/
48. http://www.valmin.org/
49. https://www.samcode.co.za/samcode-ssc/samval
50. Cooper, D., & Chapman, C. (1987). *Risk analysis for large projects – Models, methods and cases*. Chichester: Wiley.
51. Seeger, M. (1999). *MSc Eng project report: Identification, quantification and simulation of risks facing Greenfields mining projects in sub-Saharan Africa*. Johannesburg, South Africa: University of the Witwatersrand.
52. https://equator-principles.com/
53. Norton Rose Fulbright. *Finding finance in the mining and minerals sector – A guide for mine developers*. http://www.nortonrosefulbright.com/knowledge/publications/28747/finding-finance-in-the-mining-and-minerals-sector-a-guide-for-mine-developers
54. Smith, L. D. (1994, September). Checklist for economic evaluations of mineral projects. *Canadian Institute of Mining, 87*(983), 32–37.
55. Fisher, B. (2014). *Six secrets of capital raising*. San Francisco, CA: Berret–Koehler.
56. https://www.mining.com/web/bre-x-scandal-a-history-timeline/

57. SRK Consulting. (2013, August). *Preliminary economic assessment for the F2 Gold System, Phoenix Gold Project, Red Lake, Ontario*. https://s21.q4cdn.com/960886365/files/doc_presentations/Technical-report-phoenix-gold.pdf
58. *Gold streaming agreement Rubicon Minerals Corporation US$75.0 million with RGLD Gold AG*. (2014, February). https://www.sec.gov/Archives/edgar/data/1057791/000105779114000008/ex991.pdf
59. SRK Consulting. (2014, October). *NI 43-101 technical report feasibility study Mt. Hamilton gold and silver project centennial deposit and Seligman deposit*. https://solitarioxr.com/news/MtHamilton_NI%2043-101_09_20141016.pdf
60. *Net smelter return royalty agreement*. Mt Hamilton Gold Mine. (2012). https://www.lawinsider.com/contracts/447AArpdd6jZxAuNSFDjVE/solitario-exploration-royalty-corp/net-smelter-return-royalty-agreement/2012-06-13

The manufacturer's authorised representative in the EU is Springer Nature Customer Service Centre GmbH, Europaplatz 3, 69115 Heidelberg, Germany. If you have any concerns regarding our products, please contact ProductSafety@springernature.com

Printed and bound by CPI Group (UK) Ltd, Croydon, CR0 4YY
05/10/2026
02228145-0007